U0305782

城市行走书系
策划：江岱，姜庆共

上海邬达克建筑地图
策划统筹：华霞虹

中文文字：华霞虹
英文文字：乔争月
地图：齐斐然，卢恺琦，孙晓悦
绘图：华霞虹（除注明外）
摄影：席闻雷，上海市城市建设档案馆，
上海章明建筑设计事务所
英文校对：比瓦什·穆克何几

责任编辑：张翠
书籍设计：孙晓悦
助理：刘羽云，刘嘉纬

鸣谢：
郑时龄，伍江，罗小未，赵天佐，项秉仁，
卢卡·彭切里尼，勒诺·希特坎普，江似虹，
海博，海笛，李雅娟，彼得·雅诺西，
维拉格·切伊迪，左志，谈会明，
维多利亚·格雷厄姆

上海市城市建设档案馆
加拿大维多利亚大学
匈牙利驻上海总领事馆
上海章明建筑设计事务所
上海市城市规划设计研究院
上海图书馆
上海徐家汇藏书楼
上海市档案馆
上海市作家协会
复旦大学附属华东医院

CityWalk Series
Curator: Jiang Dai, Jiang Qinggong

Shanghai Hudec Architecture
Curator: Hua Xiahong

Chinese Text: Hua Xiahong
English Text: Michelle Qiao
Map: Kató Siegfried, Boglárka Lukács,
Sun Xiaoyue
Illustration: Hua Xiahong
Photograph: Xi Wenlei, Shanghai Urban
Construction Archives, Shanghai Zhangming
Architectural Design Firm
English Revision: Bivash Mukherjee

Editor: Sarah Zhang
Book Designer: Sun Xiaoyue
Assistant: Liu Yuyun, Liu Jiawei

Acknowledgements:
Zheng Shiling, Wu Jiang, Luo Xiaowei,
Zhao Tianzuo, Xiang Bingren, Luca Poncellini,
Lenore Hietkamp, Tess Johnston, Tamás Hajba,
Judit Hajba, Révész Ágota, Péter Jánossy,
Virág Csejdy, Zuo Zhi, Tan Huiming,
Victoria Graham

Shanghai Urban Construction Archives
University of Victoria, Canada
Consulate General of the Republic of
Hungary in Shanghai
Shanghai Zhangming Architectural Design Firm
Shanghai Urban Planning and Design
Research Institute
Shanghai Library
The Xujiahui Library
Shanghai Municipal Archives
Shanghai Writers' Association
Huadong Hospital Affiliated Fudan University

Shanghai Hudec Architecture

上海邬达克建筑地图

华霞虹 乔争月 [匈]齐斐然 [匈]卢恺琦 著
Hua Xiahong Michelle Qiao Kató Siegfried Boglárka Lukács

同济大学出版社
TONGJI UNIVERSITY PRESS

聯合大樓
261

目录

序言 ······ 8
上海的邬达克 ······ 11
——一位建筑师的传奇

上海邬达克建筑地图 ······ 18
生平建筑年表+地图 ······ 20

A 区 ······ 28
慕尔堂 ······ 30
大光明大戏院 ······ 36
国际饭店 ······ 42

B 区 ······ 48
美国总会 ······ 50
四行储蓄会联合大楼 ······ 54
广学大楼 ······ 58
真光大楼 ······ 58

C 区 ······ 64
巨籁达路 22 栋住宅 ······ 66
盘滕住宅 ······ 70
宏恩医院 ······ 74
爱文义公寓 ······ 80
吴同文住宅 ······ 84

D 区 ······ 90
诺曼底公寓 ······ 92
邬达克自宅 ······ 96
孙科住宅 ······ 100
交通大学工程馆 ······ 106
哥伦比亚住宅圈 ······ 110
达华公寓 ······ 116

E 区 ······ 120
何东住宅 ······ 122
爱司公寓 ······ 126
刘吉生住宅 ······ 130
斜桥弄巨厦 ······ 136
震旦女子文理学院 ······ 142

F 区 ······ 146
中西女塾社交堂 ······ 148
中西女塾景莲堂 ······ 148
上海啤酒厂 ······ 154

G 区 ······ 158
息焉堂 ······ 160

邬达克的上海 ······ 164
——这美丽的香格里拉

上海邬达克建筑不完全名录 ······ 169
延伸研究 ······ 176
相关信息 ······ 177
推荐阅读 ······ 177
图片版权 ······ 179

Table of Contents

Preface ······ 9

The Making of an Architect ······ 15

Shanghai Hudec Architecture ······ 18

- Chronology + Map ······ 20

Zone A ······ 28

- Moore Memorial Church ······ 30
- Grand Theatre ······ 36
- Park Hotel ······ 42

Zone B ······ 48

- American Club ······ 50
- Union Building of the Joint Savings Society ······ 54
- Christian Literature Society Building ······ 58
- China Baptist Publication Building ······ 58

Zone C ······ 64

- The 22 Residences on Route Ratard ······ 66
- Jean Beudin's Residence ······ 70
- Country Hospital ······ 74
- Avenue Apartments ······ 80
- D.V.Woo's Residence ······ 84

Zone D ······ 90

- Normandie Apartments ······ 92
- Hudec's Residence ······ 96
- Sun Ke's Residence ······ 100
- Engineering and Laboratory Building of Chiao Tung University ······ 106
- Columbia Circle ······ 110
- Hubertus Court ······ 116

Zone E ······ 120

- Ho Tung's Residence ······ 122
- Estrella Apartments ······ 126
- Liu Jisheng's Residence ······ 130
- P.C.Woo's Residence ······ 136
- Aurora College for Women ······ 142

Zone F ······ 146

- Social Hall, McTyeire School for Girls ······ 148
- McGregor Hall, McTyeire School for Girls ······ 148
- Union Brewery Ltd. ······ 154

Zone G ······ 158

- Sieh Yih Chapel ······ 160

The Beautiful Shangri-La ······ 166

Shanghai Hudec Architecture Incomplete Directory ······ 169

More to Explore ······ 176

References ······ 177

Recommended Readings ······ 177

Image Copyrights ······ 179

序言

上海的优秀近代建筑是由中国近代第一代建筑师以及众多的外国建筑师所共同创造的。斯裔匈籍建筑师邬达克是上海新建筑的一位先锋，他善于学习世界各国的建筑式样，孜孜以求建筑的时代精神。他的建筑风格历经新古典主义、表现主义、装饰艺术派以及现代建筑风格，仿佛建筑风格的大全，既有当时欧美建筑的直接影响，也有建筑师个人的创造。邬达克留下的大量建筑作品，书写了上海近代建筑史辉煌的一页篇章。上海培育了邬达克，而作为现代建筑的倡导者，他也创造了上海建筑的摩登风格。

邬达克于 1918 年来到上海，在美国建筑师克利开设的克利洋行工作。在此期间他与克利合作设计了诺曼底公寓、美国总会、四行储蓄会联合大楼等一系列作品。这些作品均为复古样式，但邬达克个人的某些风格，如喜爱用面砖饰面，已开始形成。

1925 年邬达克自己开业，在 20 世纪 30 年代达到他的建筑师生涯的鼎盛时期。随着国际新建筑风格的出现，邬达克的设计风格也发生了重大转变，成为上海新风格建筑最引人注目的大力推动者。他的设计风格的转变最初出现在 1932 年建成的广学大楼上，这座表现主义风格的办公建筑的立面上还留有一些传统风格的痕迹，其哥特式尖券的造型和褐色的面砖使整个建筑的造型十分简洁，又不失华丽和凝重。

具有强烈时代感的大光明大戏院于 1933 年的落成，标志着邬达克设计风格完成了彻底的转变，他的新潮设计立刻受到建筑界的广泛关注，并由此奠定了他作为上海最有影响的现代建筑师的地位。1934 年 12 月，几乎是美国 30 年代摩天楼的直接翻版的高达 83.8 米的国际饭店落成。这座大楼不仅造型新颖，融汇了现代建筑和表现主义的语言，其结构、设备都代表了当时上海甚至远东地区的最高水平，由此奠定了邬达克在上海建筑史上不可动摇的先锋地位。而 1938 年建成的吴同文住宅，设计风格更接近国际式。圆弧形的大片落地玻璃窗、强烈的水平线处理和流线形的室外大楼梯等，使这座住宅建筑成为上海现代建筑的代表作。

邬达克的建筑风格注重形象的整体几何性，造型丰富，细部处理细腻，而且总能有新颖的构思，思路从未枯竭，他所设计的 50 多项建筑中没有重复出现的母题，这点对于一位建筑师是十分难能可贵的。

本书的中文作者华霞虹博士近 14 年来一直在研究邬达克和他的建筑，国内研究者众多，无出其右。这本由她和其他学者合著的《上海邬达克建筑地图》将邬达克作品的精华，也是上海近代建筑的一部分精华极好地展示给读者。认识邬达克，也是认识上海城市的历史和未来。

2012 年 12 月 5 日

Preface

The architecture of modern times in Shanghai was co-created by the first generation of Chinese modern architects and numerous foreign architects. L.E.Hudec (1893-1958) was undoubtedly the pioneer of the new architecture in Shanghai. He was good at learning from world architecture and captured the spirit of the time. Like an encyclopedia of architecture style, his style had gone through neo-classicism, expressionism, Art Deco and modernism, which not only reflected the direct European and American influences, but also the architect's personal creativity. Hudec has left behind a lot of work that is remarkable in Shanghai's architectural history. He was fostered by Shanghai and he created the modern style for Shanghai.

Hudec came to Shanghai in 1918 and worked for American architect firm R.A. Curry. They co-designed Normandie Apartments, American Club, Union Building of the Joint Savings Society among others, which were nearly all classical revival style, but some of his personal style was formed here.

In 1925, Hudec opened his own architect company, and in 1930s reached the zenith of his career. Along with the emergence of international style, his architecture had undergone a significant transformation. He became the most remarkable protagonist of new architecture in Shanghai, which first appeared in the Christian Literature Society Building in 1932 — an expressionist office building with certain Gothic detail. Its form and brown tile made the building quite simple but also magnificent.

The Grand Theatre with a strong modern style was completed in 1933 which meant an overall transformation of Hudec's style. His new style has been paid more and more attention in architectural field, making him the most influential architect in Shanghai. In December 1934, the Park Hotel, with its height of 83.8 meters opened, which was almost a copy of the American skyscrapers of 1930s. Its new form mixed with modern and expressionist architectural language, along with its structure and mechanical engineering, represented the highest level of excellence in Shanghai and in the Far East. As one of the masterpieces of Shanghai modern architecture, D.V.Woo's residence with large curved windows and horizontal stream lines was completed in 1938. Its design captured the mainstream international style.

His main style is reflected in the integrated geometry, varied form, rich detail and original ideas. His thinking never dried out, and among his more than fifty projects, the motive has never been the same, which is indeed rare for an architect.

For the last 14 years, Dr. Hua Xiahong has devoted herself to the study of Hudec and his architecture. Among numerous researchers, she was excellent. The Shanghai Hudec Architecture has shown the essence of Hudec's projects, which is also one part of the essence of Shanghai architecture. To know Hudec, is to know the history of Shanghai and the city's future.

Zheng Shiling

5th. December, 2012

上海的邬达克

——一位建筑师的传奇

一位外国建筑师，一名逃亡的战俘，在异国他乡的29年（1918−1947）间，建成项目不下50个（单体超过100幢），其中31个项目（单体超过50幢）已先后被列为上海市优秀历史建筑，国际饭店更成为全国文物保护单位。这些工程涉及办公、旅馆、医院、教堂、影院、学校、工厂、公寓、会所、私宅等众多类型，外观包括古典主义、折中主义、装饰艺术、表现主义、现代主义和地域主义等不同风格，区位从外滩源附近蔓延到西郊乃至近吴淞口。邬达克在上海的作品数量之多、种类之全、分布之广、质量之高，在世界建筑史上也不算多见。

更不寻常的是，这位20世纪30年代上海滩炙手可热的建筑明星，70多年后再次成为大众耳熟能详的“上海建筑师”，其名字和作品频频出现在报刊、电视、展览等媒体中。“邬粉”人群之众，不仅引发了“城市暴走”等自发活动，甚至开始左右其作品的保护与更新计划。

邬达克的传奇，其内涵究竟是什么？它又是如何产生并延续至今的呢？

1 传奇的产生：时势造英雄

如果用一句话概括传奇产生的原因，那就是：邬达克在合适的时间，来到合适的地方，从事了合适的职业。

20世纪二三十年代是近代上海经济最繁荣的阶段。上海是当时亚洲最大的城市，远东最大的贸易中心、金融中心和工业中心，与纽约、伦敦、巴黎并称世界四大都市。经济活力、人口剧增、快速城市化带来房地产业和建筑业的蓬勃。

1927年（上海建市）到1937年（“二战”爆发）是上海城市和建筑发展最为鼎盛的十年，而邬达克也正是在这一时期建成了他60%的作品，其中包含质量最高的国际饭店、大光明大戏院、慕尔堂、德国新福音教堂等，而吴同文住宅也已完成设计，进入施工阶段。

2 传奇的内涵：边缘、摩登与多元

近代上海是中外建筑师云集之地，无论从工程数量、质量、影响力、创新度等各方面来看，邬达克都并非一枝独秀，而是始终面临激烈的竞争。正因如此，才有上海今天富有魅力的诸多历史文化风貌区和大批优秀建筑遗产。

然而，邬达克及其作品的确算得上近代上海都市文化中一个代表性个案。通过与当年竞争者，特别是他最强劲的对手也是同时期上海最大的外商设计机构——公和洋行进行对比，或可帮助我

们更清晰地了解这种传奇的内涵，也就是其建筑作品的独特性。

公和洋行是正统老牌子，在 20 世纪 20 年代外滩改建中占据垄断地位，代表着当时租界强势移植并占主流地位的西方文化。相反，邬达克始终处于尴尬的边缘，却也因此获益。他的国籍长期不确定，也不像英、美、法、日等国人那样有“治外法权”的庇护，邬达克洋行的法律纠纷须在中国法庭仲裁。不过，这反而使他赢得了中国业主的信任，并合力创造了最杰出的作品。

如果说外滩建筑可能残留挥之不去的“鸦片”味的话，国际饭店成为“远东第一高楼”则不仅是建筑师的骄傲，也是上海的骄傲，中国的骄傲。正是这种中西边缘的身份使邬达克的传奇得以超越时代和政治的局限。

求新求变是商业社会的生存之道，也是近代上海的文化精髓。公和洋行主要凭借古典功力称雄，后期虽逐渐转向现代，但与前期复古作品无论在材质上还是理性精神上都是统一的，项目的规模和类型也差别不大，这是稳定的大公司的实力表现。

相比之下，单枪匹马的邬达克必须周旋于各类业主之间，不仅公司规模会随行就市，设计风格也常客随主便。在克利洋行时，因业主需求多采用古典形式；自行开业后，则把采用最新技术、最新形式,利用城市复杂的基地条件“螺蛳壳里做道场”作为竞争的法宝。

同时，中国业主在租界居于弱势，邬达克作为西方建筑师也处于边缘，这种压力也很容易转化为“别苗头”的动力，比如像国际饭店这样创纪录的摩天楼，对业主和设计师同样具有广告效应。

因为业主、项目区位、类型、规模、投资各不相同，加上邬达克本人多元的文化背景和虔诚的宗教信仰，其作品的形式和材质都比同时期在上海的其他建筑师更为丰富，尤其是对哥特复兴和表现主义建筑风格的探索，在上海近代建筑史中显得格外突出。

然而，饱受战争伤害的邬达克本能地躲避政治是非，他与中国社会和文化采取相对疏离姿态，除局部室内陈设外，其作品外观不像墨菲等外国建筑师那样热衷于中国传统建筑形式。他也秉持建筑师属于服务行业的传统思想，设计的创新和多样更大程度上局限于技术和美学层面，并不具有意识形态的革命性和对社会转型的深入思考。这与同时期欧洲如火如荼的现代主义建筑运动乃至近代上海的中国建筑师都有明显的差别。

事实上，邬达克的建筑作品在中西边缘精明发展，追求摩登，多元融合的特点也是所谓“海派文化”的缩影。

3 传奇的延续：怀旧消费热

资料丰富便于研究是“邬达克热”在半个多世纪后得以延续的重要基础。一方面，20 世纪二三十年代是世界范围现代主义建筑运动和近代中国建筑行业和学科蓬勃发展的时期。除档案馆留有大量工程记录和历史图纸外，邬达克的代表作在当时中外专业和大众媒体上也有广泛报道。

另一方面，出于证明身份，也出于个人对考古的兴趣，邬达克留下了大量草图、照片、剪报、信件等历史资料，这为研究者提供了丰富的素材。

近年来中国与匈牙利、斯洛伐克等东欧国家对外交流和贸易上的需求是促进邬达克研究的另一个动力。

中国文化消费市场的巨大缺口，城市更新过程中对历史建筑的开发利用以及由此带来的可观的经济、社会和文化效益，上海建设创意之都、全球城市的努力，这些都是所谓“上个世纪黄金年代”的怀旧消费热的基础。

经历了八九十年世事沧桑的大批老房子本身充满故事性，加上建筑师的人生经历跌宕起伏富有戏剧性，“邬达克的传奇”不断延续。

一个人的命运不可思议地契合了一座城市的命运，一个人的才能与抱负不失时机地响应了一座城市的梦想与追求，这就是一位建筑师能在遥远他乡取得如此瞩目的成就和影响的最终谜底。

邬达克的传奇，也是上海都市文化的传奇。

图书在版编目（CIP）数据

上海邬达克建筑地图 / 华霞虹等著. -- 上海：同济大学出版社, 2013.1（2024.10重印）

ISBN 978-7-5608-5061-0

Ⅰ. ①上… Ⅱ. ①华… Ⅲ. ①建筑设计－研究－上海市②上海邬（1893～1958）－建筑设计－研究 Ⅳ. ①TU2

中国版本图书馆CIP数据核字(2012)第300848号

上海邬达克建筑地图

华霞虹　乔争月　[匈]齐斐然　[匈]卢恺琦　著

出 品 人：支文军
责任编辑：张　翠
责任校对：徐春莲
出版发行：同济大学出版社 www.tongjipress.com.cn
地　　址：上海市四平路1239号　邮编：200092
电　　话：021－65985622
经　　销：全国新华书店
印　　刷：上海雅昌艺术印刷有限公司
开　　本：787mm×1092mm　1/36
印　　张：5
印　　数：34 101－37 200
字　　数：150 000
版　　次：2013年1月第1版
印　　次：2024年10月第12次印刷
书　　号：ISBN 978－7－5608－5061－0
定　　价：56.00元

图片版权

摄影

席闻雷 P4、P6、P10、P14、P18-19、P34-35、P42、P50、P54、P58、P60、P63、P66、P68、P70、P74、P84、P89、P92、P106、P110、P112、P115、P116、P122、P126、P130、P134、P135、P136、P142-143、P148-149、P150、P152、P154、P168、P179

上海市城市建设档案馆 P22（左）、P23（右）、P30、P79（上）、P80、P96、P100-101、P102、P138、P160、P174（46）

上海章明建筑设计事务所 P3、P36、P40-41

加拿大维多利亚大学图书馆 P21、P22（右）、P23（左）、P24、P169、P170、P171、P172（22、23、25、34）、P174（40、53）

华霞虹 P25、P172（35）、P174（45）

华东医院 P79（下）

图纸

《建筑月刊》第二卷第三号（1933）P44、P46、P47
《建筑月刊》第三卷第一号（1934）P140、P141
工部局编《中国建筑师和营造商年鉴 1925》P80
上海图书馆馆藏
《上海啤酒公司 25 周年纪念特刊》P156
《上海泰晤士报》（星期刊）工业版，
1930 年圣诞专刊增刊 P162

华霞虹根据以下历史资料重新手绘：
上海市城市建设档案馆馆藏图纸
P32、P38、P52、P56、P62、P82、P86、P88、P98、P104、P118、P128、P134
上海民用建筑设计研究院编
《上海公寓建筑图集》（1985）P94
维多利亚大学图书馆 P108、P114
hudecproject.com P74、P126

Image Copyrights

Photos

Xi Wenlei P4, P6, P10, P14, P18-19, P34-35, P42, P50, P54, P58, P60, P63, P66, P68, P70, P74, P84, P89, P92, P106, P110, P112, P115, P116, P122, P126, P130, P134, P135, P136, P142-143, P148-149, P150, P152, P154, P168, P179

Shanghai Urban Construction Archives P22(left), P23(right), P30, P79(up), P80, P96, P100-101, P102, P138, P160, P174(46)

Shanghai Zhangming Architectural Design Firm P3, P36, P40-41

University of Victoria, Canada P21, P22(right), P23(left), P24, P169, P170, P171, P172(22, 23, 25, 34), P174(40, 53)

Hua Xiahong P25, P172(35), P174(45)

Huadong Hospital P79(down)

Plans

The Builder Magazine (No. 3, Volume 2, 1933) P44, P46, P47
The Builder Magazine (No. 1, Volume 3, 1934) P140, P141
J. T. W. Brooke & R. W. Pavis eds. *The China Architects and Builders Compendium 1925* P80
Collections of Shanghai Library, *25th Anniversary of Union Brewery Ltd. Shanghai , Special Issue* P156
The Shanghai Sunday Times, Industrial Section, Supplement to Special Xmas Issue, 1930 P162

Reproduced by Hua xiahong based on following documents:
Collections of Shanghai Urban Construction Archives P32, P38, P52, P56, P62, P82, P86, P88, P98, P104, P118, P128, P134
The Plan Atlas of Shanghai Apartment Buildings (1985), produced by Shanghai Civil Architecture Design and Research Institute P94
University of Victoria library, Canada P108, P114
hudecproject.com P74, P126

相关信息
References

专著及论文 Book & Thesis

1 Lenore Hietkamp, *Laszlo Hudec and The Park Hotel in Shanghai*[M], China Diamond River Books, 2012
2 Luca Poncellini & Júlia Csejdy, *László Hudec*[M], Holnap Kiadó, 2010
3 Jánossy Péter Sámuel & Deke Erh, *Life and Work of László Hudec*[M], Építésügyi Tájékozta tási Központ Kft., 2010
4 上海市城市规划管理局、上海市城市建设档案馆编，上海邬达克建筑 [M]，上海科学普及出版社，2008
5 上海市城市规划研究院 / 上海现代建筑设计集团 / 同济大学建筑与城市规划学院编著，绿房子 [M]，同济大学出版社，2014
6 本书编委会编，邬达克的家——番禺路 129 号的前世今生，上海远东出版社，2015
7 LIU Bingkun, *Laszlo E. Hudec and Modern Architecture in Shanghai*, 1918-1937 [D]. Hongkong University, 2005
8 华霞虹 . 邬达克在上海作品的评析 [D]. 同济大学：2000

网站 Website

1 加拿大维多利亚大学图书馆特别收藏——邬达克档案 Laszlo Hudec Fonds http://library.uvic.ca/site/spcoll/guides/sc132.html
2 邬达克年网站 www.hudec.sh
3 邬达克遗产项目网站 www.hudecproject .com
4 豆瓣网“邬达克的老房子”社区 www.douban.com/group/hudec/
5 新浪博客“邬达克 L.E. Hudec” http://blog.sina.com.cn/tthmbdbk

纪录片 Documentary

1 Ladislav Kabos, *The Man Who Changed Shanghai*, www.ladislavhudec.eu
2 Réka Pigniczky, *The life of László Hudec*, 2010/6, www.56films.com

推荐阅读
Recommended Readings

邬达克，[意] 卢卡 · 彭切里尼、[匈] 尤利娅 · 切伊迪著，华霞虹、乔争月译，同济大学出版社，2013
上海近代建筑风格，郑时龄著，上海教育出版社，1999
上海百年建筑史：1840-1949（第二版），伍江著，同济大学出版社，2008
上海建筑指南，罗小未主编，上海人民美术出版社，1996
上海近代建筑史稿，陈从周、章明著，上海三联书店，1988
上海近代城市建筑，王绍周著，江苏科学技术出版社，1989
A Last look: Western Architecture in Old Shanghai by Tess Johnston & Deke Erh, Old China Hand Press,1993
中国近代建筑史（1-5），赖德霖、伍江、徐苏斌主编，中国建筑工业出版社，2016
“外滩源”研究：上海原英领馆街区及其建筑的时空变迁 (1843-1937)，王方著，东南大学出版社，2011
上海武康路：风貌保护道路的历史研究与保护规划探索，沙永杰、纪雁、钱宗灏著，同济大学出版社，2009
Art Deco 的源与流：中西摩登建筑关系研究，许乙弘著，东南大学出版社，2006
上海百年建筑师和营造师，娄承浩、薛顺生编著，同济大学出版社，2011
大光明 · 光影八十年，上海大光明电影院有限公司编，同济大学出版社，2009
近代上海城市公共空间（1843-1949），王敏、魏兵兵、江文君、邵建著，上海辞书出版社，2011
上海摩登（修订版）：一种新都市文化在中国（1930-1945），李欧梵著，毛尖译，人民文学出版社，2010

GRAND
THEATRE
上島
上島咖啡
千鍋居

31 息焉堂（1929-1931）
今西郊天主堂
可乐路 1 号（近哈密路）
长宁区

32 慕尔堂（1926-1931）
今沐恩堂
西藏中路 316 号（近汉口路）
黄浦区

33 交通大学工程馆（1931）
华山路 1954 号（近番禺路）
徐汇区

34 哥伦比亚住宅圈（1928-1932）
今外国弄堂私人住宅
新华路 119、155、185、211、236、248、276、294、329 弄部分住宅（近番禺路）
长宁区

35 广学大楼（1930-1932）
虎丘路 128 号（近香港路）
黄浦区

36 真光大楼（1930-1932）
圆明园路 209 号（近香港路）
黄浦区

37 德国新福音教堂（1930-1932）
已拆除，原址建上海希尔顿酒店和国际贵都大饭店
延安中路华山路转角
静安区

38 国际礼拜堂牧师住宅（1932）
今徐汇区体育局
乌鲁木齐南路 64 号
徐汇区

39 爱文义公寓（1931-1932）
今联华公寓
北京西路 1341-1383 号（近铜仁路）
静安区

40 辣斐路花园住宅（1931-1932）
今花园住宅
复兴西路 133 号
徐汇区

31 Sieh Yih Chapel (1929-1931)
now Catholic Country Church
No. 1 Kele Rd (near Hami Rd)
Changning District

32 Moore Memorial Church (1926-1931)
now Mu'en Church
No. 316 Middle Xizang Rd (near Hankou Rd)
Huangpu District

33 Engineering and Laboratory Building of Chiao Tung University (1931)
No. 1954 Huashan Rd (near Panyu Rd)
Xuhui District

34 Columbia Circle (1928-1932)
now Foreigner's Lane Garden Villa
Some Villas of Lane 119, 155, 185, 211, 236, 248, 276, 294, 329 Xinhua Rd (near Panyu Rd)
Changning District

35 Christian Literature Society Building (1930-1932)
No. 128 Huqiu Rd (near Xianggang Rd)
Huangpu District

36 China Baptist Publication Building (1930-1932)
No. 209 Yuanmingyuan Rd (near Xianggang Rd)
Huangpu District

37 New German Lutheran Church (1930-1932)
now demolished, same site of Hilton Hotel and Equatarial Shanghai,
Corner of Middle Yan'an Rd, Huashan Rd
Jing'an District

38 Manse for Community Church (1932)
now Xuhui District Sports Council
No. 64 South Wulumuqi Rd
Xuhui District

39 Avenue Apartments (1931-1932)
now Lianhua Apartments
No. 1341-1383 West Beijing Rd (near Tongren Rd)
Jing'an District

40 Garden Villa on Rue Lafayette (1931-1932)
now Garden Villa
No. 133 West Fuxing Rd
Xuhui District

41 斜桥弄巨厦（1931-1932）
今上海市公惠医院
石门一路 315 弄 6 号（近南京西路）
静安区

42 大光明大戏院（1931-1933）
今大光明电影院
南京西路 216 号（近黄河路）
黄浦区

43 辣斐大戏院（1932-1933）
今拉法耶艺术设计中心
复兴中路 323 号
黄浦区

44 上海啤酒厂（1933-1934）
今苏州河梦清园
宜昌路 130 号（近中潭路）
普陀区

45 国际饭店（1931-1934）
南京西路 170 号（近黄河路）
黄浦区

46 中西女塾景莲堂（1921-1935）
今市三女中五四大楼
江苏路 155 号（近武定西路）
长宁区

47 上海朝阳路圣心女子职业学校（1936）
今长城饭店
眉州路 272 号（近沈阳路）
杨浦区

48 达华公寓（1935-1937）
今上海锦江都城达华酒店
延安西路 918 号（近江苏路）
长宁区

41 P.C.Woo's Residence (1931-1932)
now Shanghai Gonghui Hospital
No. 6 Lane 315 Shimen 1. Rd
(near West Nanjing Rd)
Jing'an District

42 Grand Theatre (1931-1933)
now Grand Cinema
No. 216 West Nanjing Rd
(near Huanghe Rd),
Huangpu District

43 Lafayette Cinema (1932-1933)
now Lafayette Arts & Design Center
No. 323 Middle Fuxing Rd
Huangpu District

44 Union Brewery Ltd. (1933-1934)
now Mengqing Park
No. 130 Yichang Rd (near Zhongtan Rd)
Putuo District

45 Park Hotel (1931-1934)
No. 170 West Nanjing Rd (near Huanghe Rd)
Huangpu District

46 McGregor Hall, McTyeire School for Girls (1921-1935)
now May 4 Building, Shanghai No.3 Girls High School
No. 155 Jiangsu Rd (near West Wuding Rd)
Changning District

47 Sacred Heart Vocational College for Girls (1936)
now Great Wall Hotel
No.272 Meizhou Rd (near Shenyang Rd)
Yangpu District

48 Hubertus Court (1935-1937)
now Shanghai Jinjiang Metropolo Dahua Hotel
No. 918 West Yan'an Rd (near Jiangsu Rd)
Changning District

43 47 54

49 吴同文住宅（1935-1938）
今上海规划师之家和上海城市规划博物馆（改建中）
铜仁路 333 号（近北京西路）
静安区

50 震旦女子文理学院附属圣心女子小学（1938）
原向明高级中学，已拆除
长乐路 141 号（近瑞金一路）
黄浦区

51 震旦女子文理学院（1937-1939）
今向明高级中学震旦楼
长乐路 141 号（近瑞金一路）
黄埔区

52 俄罗斯天主学校男童宿舍（1941）
今向明高级中学国际部（晓光楼）
长乐路 141 号（近瑞金一路）
黄浦区

53 意大利总会新大礼堂（1941）
已拆除
原静安寺路
静安区

54 大西路德国学校图书馆（1941）
已拆除
原大西路 7 号
静安区

49 D.V.Woo's Residence (1935-1938)
now Shanghai Urban Planner's Saloon and Urban Planning Museum (under construction)
No. 333 Tongren Rd (near West Beijing Rd)
Jing'an District

50 Chinese School of the Convent of the Sacred Heart (1938)
former Xiangming Senior High School, demolished
No. 141 Changle Rd (near Ruijin 1. Rd)
Huangpu District

51 Aurora College for Women (1937-1939)
now Xiangming Senior High School Aurora Building
No. 141 Changle Rd (near Ruijin 1. Rd)
Huangpu District

52 Russian Catholic School Hostel for Boys (1941)
now International Department of Xiangming Senior High School (Xiaoguang Building)
No.141 Changle Road (near Ruijin 1. Rd)
Huangpu District

53 New Auditorium for Italian O.N.D Club (1941)
demolished
former Bubbling Well Rd
Jing'an District

54 German School Laboratory (1941)
demolished
former No.7 Great Western Rd
Jing'an District

延伸研究

More to Explore

研究历史就像串联散落已久的珍珠，总能发现还有遗漏和错装的宝贝，这既是遗憾，也是趣味所在。关于邬达克在上海的建筑故事，我们还有很多可以去发掘，比如：

1. 邬达克洋行鼎盛时期（1930-1935）至少有 15 位中国建筑师，他们是谁？起了什么作用？后来去向如何？

2. 匈牙利邬达克文化基金会存有一本当年邬达克寄给的作品集，其中收录了从 1918-1924 年间 9 个住宅项目的蓝图（无签名）和室内外照片，备注是：邬达克建筑师在上海设计的住宅作品。其中包括古典主义的独立住宅，依次为：何东住宅、盘滕住宅、杜克住宅、马迪耶住宅、霍肯多夫住宅、梅里霭住宅和耶斯佩森住宅，还有巨鹿路 22 栋住宅和宝建路 9 栋住宅。其中缺乏进一步信息的是杜克住宅耶斯佩森住宅和宝建路 9 栋住宅。按时间推断这是邬达克在克里洋行时期参与过的项目，图纸上没有签名是因为邬达克当时没有资格吗？还是邬达克在开设自己的事务所时把并非自己设计的项目也列入其中以增强业绩背景？

3. 宝建路 9 栋住宅是否就是桃江路 7，15，21，25 和东平路 2，4，6，8 一批住宅？因桃江路和宝庆路即原宝建路，而这些住宅的形式跟邬达克作品集中的历史照片有极高的相似度。此外，它们跟巨鹿路 22 栋住宅在入口细部上也很相似，这两个项目是统一由万国储蓄会开发的吗？

4. 在息焉堂的工程中，邬达克和潘世义究竟是什么关系？

5. 哥伦比亚住宅圈哪些花园洋房确定是邬达克设计并留存至今的？“蛋糕房”（新华路 329 弄 36 号）真的是邬达克作品吗？

6. 《建筑月刊》第二卷第三号（1933 年）“邬达克建筑师小传”中提到邬达克自行开业后承担的“虹口大戏院改装工程”是否就是新中国首座对外开放的营业性电影院虹口大戏院（该项目也得到了工部局 1941 年一份材料的证实）？

7. 今比利时驻沪总领事馆（武夷路 127 号）是邬达克为挪威商人 Frithjof Gustav Cark Hohnke 在吕西纳路（今利西路）设计的私人住宅吗？

……

Historical research is like stringing pearls together. There are a lot of missing pearls always and quite often the entire effort seems disjointed. But that is the joy of the work. There is more to Laszlo Hudec's Shanghai adventure like the following unanswered questions.

1. Hudec's own firm (L.E.Hudec Architect) employed at least 15 Chinese architects during 1930 to 1935. Who were they and what kind of roles did they perform? Whatever happened to them afterwards?

2. Hudec Cultural Foundation Hungary keeps a portfolio of 9 residential programs designed by L.E Hudec in Shanghai from 1918 to 1924, including blueprint drawings without signatures and interior and exterior photos. Some are classical style garden villas, such as Katz Residence, J. BeudinResidence, G. E. Tucker Residence, H, Madier Residence, Huckendoff Residence, Marriman Residence and Jespersens Residence. Some are real estate products like 22 houses on 9 houses on Pottier Road. According to the design and built years, they must be projects Hudec completed in R.A Curry's office. Some of them have got multiple supports that they are hudec's design while others havn't. Are they all Hudec's design? Why there is no Hudec's signature? Is it because he has not yet got the qualification or he has included in projects from someone else to convince potential clients at the early stage of his own practice?

3. Since the former Route Pottier is now Baoqing Road and Taojiang Road, is the 9 houses on Route Pottier are now No. 7,15,21,25 of Taojiang Road and No. 2,4,6,8 of Tianping Road, which look very much like the historical pictures in Hudec's portfolio? Moreover, the entrance details resemble that of the 22 houses on Route Ratard. Are they from the same developer, International Saving Society?

4. What was the relationship between Hudec and Chinese architect Pan Shiyi when they were working on the Sieh Yih Chapel?

5. Of all the villas in the Columbia Circle, which are the real Hudec works? Is the famous villa shaped like a cake also his brainchild too?

6. *The Builder* Magazine (No. 3, Volume 2, 1933) mentions Hudec had been responsible for the renovation project of Hongkou Theatre after he started his own business. (His role in the project was also proven in a 1941 Shanghai Municipal Council archive). Is it the famous Hongkou Grand Theatre (formerly on 388 Zhapu Road, demolished in 1997), China's first commercial cinema?

7. Did Hudec design the private residence for Norwegian merchant Frithjof Gustav Cark Hohnke on No. 127 Wuyi Road which now houses the Belgium Consulate General?

The Making of an Architect

László Hudec (László Edvard Hudec, or Ladislaus Edward Hudec) can only be described as a legend. As one of foreign architects who fled his native Austro-Hungary country during troubled times, he ended up stamping his class on more than 50 projects, including more than 100 buildings, during his 29-year (1918 to 1947) stay in a city far away from home. Among them, 25 projects have been listed as Shanghai Excellent Historical Buildings. His signature work, the Park Hotel, is counted as a national heritage.

His works have been varied — from offices, hotels, hospitals, churches, cinemas, schools, factories, apartments to private residences that were a kaleidoscope of styles including classicism, eclecticism, expressionism, Art Deco, modernism, regionalism, etc. The buildings are located all around the city, from the Bund area to western suburbs and even in northern Shanghai's Wu Song Kou, the confluence of the Huangpu and Yangtze Rivers. The sheer quantity, quality and variety of Hudec's works were a rarity even by world standards, especially in the history of architecture.

Over seven decades later, Hudec continues to hold sway over the intelligentsia as he did when he lived in Shanghai in the 1930s, frequently referred to in newspaper articles, in TV programs and at exhibitions. He has a devoted group of followers who passionately research, preserve and renovate his architectural relics.

How did Hudec come to enjoy his legendary status in a foreign land, especially as he arrived with almost nothing in his pocket? Why does he continue to attract new followers even in the 21th century? Below are possibly some of the reasons.

Right Time, Right City

A hero is nothing but a product of his time. Hudec, it seems, was at the right place and at the right time and was engaged in the right occupation.

Shanghai was a booming city in the 1920s and 1930s. As the largest Asian city of its time, Shanghai was also the biggest center for trade, finance and industry in the Far East and one of the four metropolitan cities along with New York, London and Paris. The dynamic economy, the soaring population and the speedy urbanization had led to a boom in real estate and construction.

The period especially between 1927 to 1937 was "a golden decade" for Shanghai's urban and architectural development, during which time Hudec had completed 60 percent of his works, including his most famous Park Hotel, the Grand Theatre and Moore Memorial Church among others.

Marginalized, Modern and Diversified

In the early last century, Shanghai attracted a galaxy of talents from around the world, including architects. Hudec was one of them.

Among his early rival and competitor was a company called Palmer & Turner, which was founded by William Salway in 1868 in Hong Kong and entered Shanghai in 1912 by building the classic Union Building (Three on the Bund).

Within decades the company played a leading role in the reconstruction of the Bund in the 1920s and designed nine of the 23 waterfront buildings such as the HSBC Building and the Sassoon House. Palmer & Turner almost became a symbol of the dominant western culture in the settlement.

Hudec, on the other hand, was in a rather unfortunate situation because of his ambiguous nationality. Unlike other expatriates in Shanghai, he could not be protected by extraterritoriality since his home country Austro-Hungary dissolved before the military defeat on the Italian front of the World War I. Lawsuit involving his company could only be solved in a Chinese court.

But it is said that when one door shuts on us another invariably opens up. Hudec benefited from his disadvantage as he won the trust of his Chinese clients, with whom he ended up creating his best works.

Most of the buildings on the Bund were built by foreign companies. But the Park Hotel, once the tallest building in the Far East, was a source of pride for Shanghai and China. It was owned by the Chinese and built by local constructors with home materials. Thus Hudec's marginalized role among the foreigners in the city actually led to his success.

The pursuit for the new and trendy was the essence of Shanghai's spirit during Hudec's time. Palmer & Turner was a firm that specialized in classic architecture. Despite their later approach to go with the modern, their works were still heavily influenced by their early classic pieces in terms of the materials they used, the scales and the genres. All were products of a big, comfortably-running firm.

In contrast, Hudec worked on his own and had the ability and flexibility to adapt to different clients.

During his early days working with the American R. A. Curry's firm, he practiced classic styles according to market demands. As soon as he ventured out on his own, he put into use the newest techniques and trendiest art forms. One of the reasons for his success over his closest competitors was his superb ability to make full use of the complicated, limited and often awkward bases.

Hudec and his Chinese clients came up with some stunning creations working against all the odds. The soaring Park Hotel was the perfect advertisement for all of them.

Compared to other architects who were working

in Shanghai around the same period, Hudec's works were amazingly diverse, a result probably of his different clients and his own cultural background. His exploration of Gothic revival style and expressionism was also noteworthy in the architectural history of Shanghai.

A victim of the wars, Hudec had cautiously avoided political pitfalls that come with them. He even maintained a distance to the Chinese society and its culture. Except for some interior details, the appearances of his buildings showed no Chinese characters, which differed from some foreign architects like Henry Murphy who loved to adapt traditional Chinese architectural forms to modern buildings.

In addition, Hudec always respected the needs of his clients. His innovative and diversified styles were mainly to serve his clients rather than lead a revolution on the conceptions, or cast a deep thought on social transition. It also differed from the modernist movement popular in Europe and Chinese architects active in Shanghai at the same time.

But Hudec's smart approach to modernism and diversification, between Chinese and western cultures, was the spirit behind Shanghai culture.

Hudec's legacy

Fortunately, Hudec left behind a rich archive of materials which is probably why he lives among the new generation of architects even today. In addition to original blueprints stored in the archives, Hudec's works have been extensively covered and debated among by professionals in the media around the world. On the other hand, he himself had plenty of drawings, photographs, newspaper clippings and letters that proved his own interest in archaeology.

The growing economic and cultural exchange between China and East European countries including Hungary and Slovakia in recent years further enhanced the study on Hudec.

Moreover, urban history and renovation of historical buildings in Chinese cities have attracted wide attentions lately. Shanghai also aims to build a city of inspiration for the future. The Hudec legacy is lasting also because people are enamored of those nearly century-old buildings created by a foreign architect whose life was full of turns and twists.

Coincidentally the destiny of one man had merged well with the destiny of one city. So his dream and capabilities echoed with the dream and pursuit of the city. That probably provides the answer to the question: how did an architect make such an impression in a city thousands of miles away from home?

Hudec's legacy is inexplicably intertwined with the Shanghai urban history.

SAMSUNG

上海邬达克建筑地图
Shanghai Hudec Architecture
CapitaLand

1893.1.8

出生于奥匈帝国的拜斯泰采巴尼亚（今斯洛伐克的班斯卡 - 比斯特里察），父亲是营造商，母亲是路德教会牧师之女

Born in Besztercebánya (now Banska Bistrica in Slovakia), in the Austro-Hungarian Monarchy, son of a master builder and the Lutheran minister's daughter

1902-1910

9 岁暑假开始在父亲建筑工地上打工；进入大学前已获木匠、泥水匠和石匠证书

Starting to work at his father's construction sites when he was 9 years old, during the summer holidays; recieved a carpenter, a mason and a stonecutter certificate before entering university

1910-1914

在布达佩斯的匈牙利皇家约瑟夫技术大学建筑系攻读学位

Studied in the department of architecture at the Hungarian Royal Joseph Technical University in Budapest

1914-1918

应征入伍，加入“一战”；1916 年于俄罗斯前线被俘，送往西伯利亚战俘集中营；后逃亡

Enlisted at the time of the outbreak of World War I; captured by Cossack patrols in the frontier in 1916; transported to prisoner's camp in Siberia; later escaped

生平 Life　1893　1902　1910　1914　1918

建筑 Architecture

拉斯洛 · 邬达克
László Hudec
1893-1958

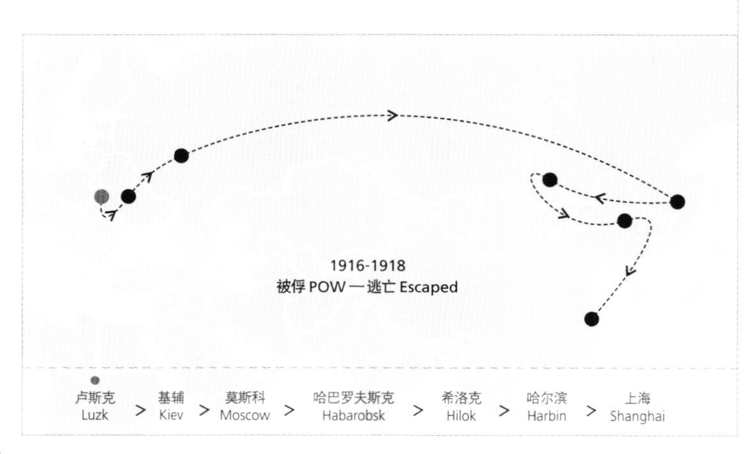

1918

经哈尔滨流亡到上海；加入克利洋行，成为助理建筑师

Arrived in Shanghai through Harbin; received employment in the office of the American architect Rowland A.Curry as a draftsman

1920、1921

父亲因病去世；到上海后第一次回欧洲处理家庭事务并旅行

Hudec's father died of a heart attack; in order to settle family affairs, travelled back to Europe for the first time after his arrival in Shanghai

1922

跟德国富商的漂亮女儿吉塞拉·迈尔结婚；着手设计和兴建位于吕西纳路的第一幢自宅

Got married to Gisela Meyer, the beautiful daughter of a wealthy German merchant; starting to build his first residence on Lucerne Road

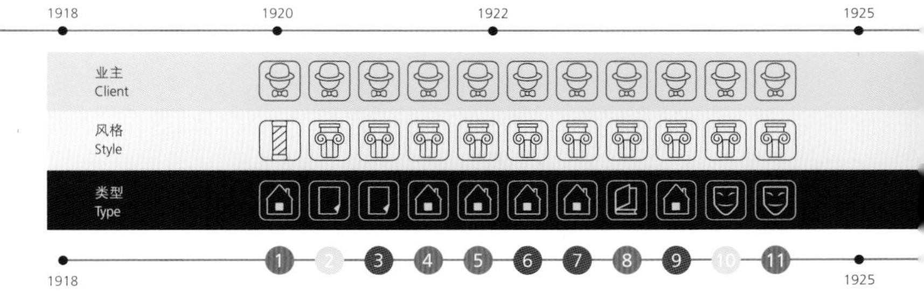

1 巨籁达路22栋住宅(1919-1920)**
2 中华懋业银行上海分行(1919-1920)
3 美丰银行(1920)**
4 何东住宅(1919-1920)**
5 盘滕住宅(1919-1920)
6 梅里霭(曼)住宅(1921)
7 霍肯多夫住宅(1921)**
8 中西女塾社交堂(1921-1922)**
9 马迪耶住宅(1922)**
10 卡尔登剧院(1923)
11 美国总会(1921-1922)**

1 The 22 Residences on Route Ratard (1919-1920)**
2 Chinese-American Bank of Commerce (1919-1920)
3 American Oriental Banking Corporation (1920)**
4 Ho Tung's Residence (1919-1920)**
5 Jean Beudin's Residence (1919-1920)
6 Meyrier or Merriman Residence (1921)
7 Huckendoff Residence (1921)**
8 Social Hall, McTyeire School for Girls (1921-1922)**
9 Madier Residence (1922)**
10 Carlton Theatre (1923)
11 American Club (1921-1922)**

1925

在横滨正金银行大厦成立事务所，开始独立执业

Opened his own architectural office, in the Building of Yokohama Specie Bank, starting his independent career

1929

赴美旅行，从纽约到旧金山，熟悉最新的摩天楼；其时正着手国际饭店项目

Travelled to USA from New York to San Francisco in order to get acquainted with the latest developments of high-rise architecture, as he was about to start the project for the Park Hotel

1925 1929 1930

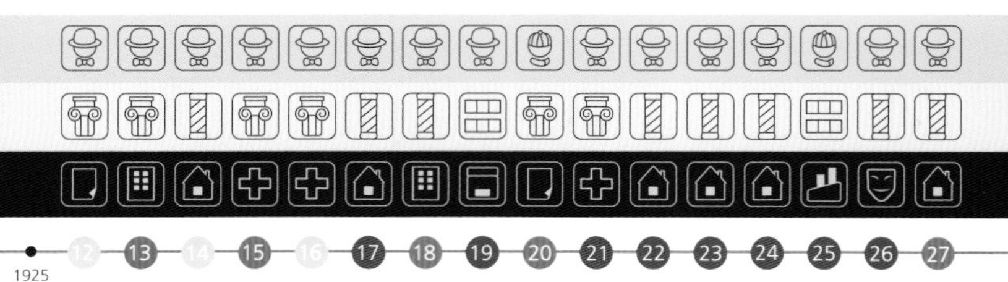

12 方西马大楼 (1924-1925)
13 诺曼底公寓 (1923-1926)**
14 邬达克首座自宅 (1922-1926)
15 宏恩医院 (1923-1926)**
16 宝隆医院 (1925-1926)
17 普益地产公司巨福路花园住宅 (1925-1926)
18 爱司公寓 (1926-1927)**
19 派克路机动车库 (1927)
20 四行储蓄会联合大楼 (1926-1928)**
21 西门外妇孺医院 (1926-1928)
22 普益地产公司西爱咸斯路花园住宅 (1925-1930)**
23 西爱咸斯路外国人私宅 (1929-1930)
24 德利那齐宅 (1929-1930)**
25 闸北电厂 (1930)
26 浙江电影院 (1929-1930)

12 Foncim Building (1924-1925)
13 Normandie Apartments (1923-1926)**
14 Hudec's 1st Residence (1922-1926)
15 Country Hospital (1923-1926) **
16 Paulun Hospital (1925-1926)
17 Garden Villas on Route Dufour for Asia Realty Co. (1925-1926)
18 Estrella Apartments (1926-1927)**
19 New Garage & Service Station of Messrs. Honigsberg Co. (1927)
20 Union Building of the Joint Savings Society (1926-1928)**
21 Margaret Williamson Hospital (1926-1928)
22 Garden Villas on Route Sieyes for Asia Realty Co. (1925-1930)**
23 H. Vladimiroff Residence on Route Sieyes (1929-1930)
24 D. Tirinnanzi Residence on Route Ferguson (1929-1930)**
25 Chapei Power Station (1930)
26 Chekiang Cinema (1929-1930)

1930

随着三个子女相继出生，出售旧宅，开始在哥伦比亚路设计并建造新的自宅和大片花园

Sold the old residence to build a larger home with a big garden close to the Columbia Circle as his three children arrived one after each other

1932

将事务所迁至自己设计的圆明园路真光大楼顶层

Moved his office to the upper floors of the "True Light" building in Yuen Ming Yuen Road, that he himself had designed

1930 1932 1935

28 29 30 31 32 33 34 35 36 37 38 39 40 41 42 43

1935

27 邬达克自宅 (1930)**
28 虹桥路雷文住宅 (1930)**
29 刘吉生住宅 (1926-1931)**
30 孙科住宅 (1929-1931)**
31 息焉堂 (1929-1931)**
32 慕尔堂 (1926-1931)**
33 交通大学工程馆 (1931)**
34 哥伦比亚住宅圈 (1928-1932)**
35 广学大楼 (1930-1932)**
36 真光大楼 (1930-1932)**
37 德国新福音教堂 (1930-1932)
38 国际礼拜堂牧师住宅 (1932)*
39 爱文义公寓 (1931-1932)**
40 辣斐路花园住宅 (1931-1932)
41 斜桥弄巨厦 (1931-1932)**
42 大光明大戏院 (1931-1933)**
43 辣斐大戏院 (1932-1933)

27 Hudec's Residence (1930)**
28 Cottage for Frank Raven (1930)**
29 Liu Jisheng's Residence (1926-1931)**
30 Sun Ke's Residence (1929-1931)**
31 Sieh Yih Chapel (1929-1931)**
32 Moore Memorial Church (1926-1931)**
33 Engineering and Laboratory Building of Chiao Tung University (1931)**
34 Columbia Circle (1928-1932)**
35 Christian Literature Society Building (1930-1932)**
36 China Baptist Publication Building (1930-1932)**
37 New German Lutheran Church (1930-1932)
38 Manse for Community Church (1932)*
39 Avenue Apartments (1931-1932)**
40 Garden Villa on Rue Lafayette (1931-1932)
41 P.C.Woo's Residence (1931-1932)**
42 Grand Theatre (1931-1933)**
43 Lafayette Cinema (1932-1933)

1937

卖掉哥伦比亚路自宅，举家搬到新建成的达华公寓底层

Sold the residence in Columbia Road, settled with his family in an apartment located on the ground floor of Hubertus Court building

1942

被任命为匈牙利领事馆驻沪荣誉领事，但不承担政治责任

Was appointed Honorary Consul of Hungary in Shanghai, his charge would not imply political duties

1947

年初，带着家人匆忙离开上海前往欧洲

Left Shanghai hastily with his family for Europe in January

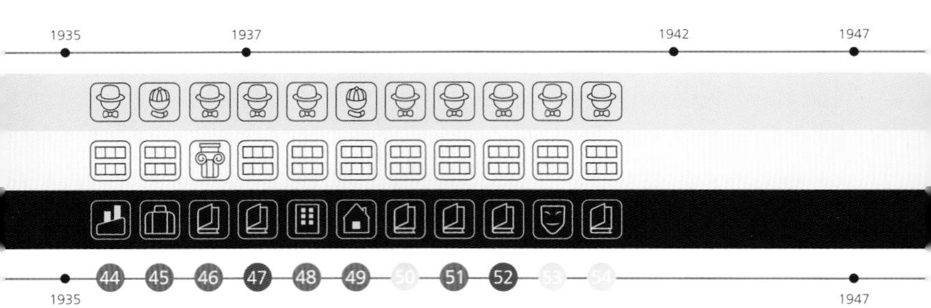

44 上海啤酒厂 (1933-1934)**
45 国际饭店 (1931-1934)***
46 中西女塾景莲堂 (1921-1935)**
47 上海朝阳路圣心女子职业学校 (1936)
48 达华公寓 (1935-1937)**
49 吴同文住宅 (1935-1938)**
50 震旦女子文理学院附属圣心女子小学 (1938)
51 震旦女子文理学院 (1937-1939)
52 俄罗斯天主学校男童宿舍 (1941)
53 意大利总会新大礼堂 (1941)
54 大西路德国学校图书馆 (1941)

44 Union Brewery Ltd. (1933-1934)**
45 Park Hotel (1931-1934)***
46 McGregor Hall, McTyeire School for Girls (1921-1935)**
47 Sacred Heart Vocational College for Girls (1936)
48 Hubertus Court (1935-1937)**
49 D.V.Woo's Residence (1935-1938)**
50 Chinese School of the Convent of the Sacred Heart (1938)
51 Aurora College for Women (1937-1939)
52 Russian Catholic School Hostel for Boys (1941)
53 New Auditorium for Italian O.N.D Club (1941)
54 German School Laboratory (1941)

n 现存+详细介绍 Existent with More Details
n 现存 Existent
n 已拆除 Demolished

*** 全国重点文物保护单位 National Heritage Buildings for Preservation
** 上海市优秀历史建筑 Shanghai Excellent Historical Buildings
* 区级优秀历史建筑 District Excellent Historical Buildings

1947

暂居瑞士，在意大利参与旨在发掘罗马凡蒂冈教堂地下圣彼得墓的秘密考古工作

Moved to Switzerland and later participated in the excavation of Saint Peter's tomb underneath the Vatican Basilica in Rome

1948

定居美国加州伯克利大学；告别建筑实践，潜心宗教和考古

Settled in Berkeley, University of California in U.S.A; retired from architectural practice and dedicated himself to religious and archeological studies

1958.10.26

在伯克利家中因心脏病突发而去世，终年 65 岁；按其遗愿，归葬故土

Died of a heart attack in his house in Berkeley at age of 65; his wife respected his will to send him back to his family's vault in hometown Besztercebánya

1947 1948 1958

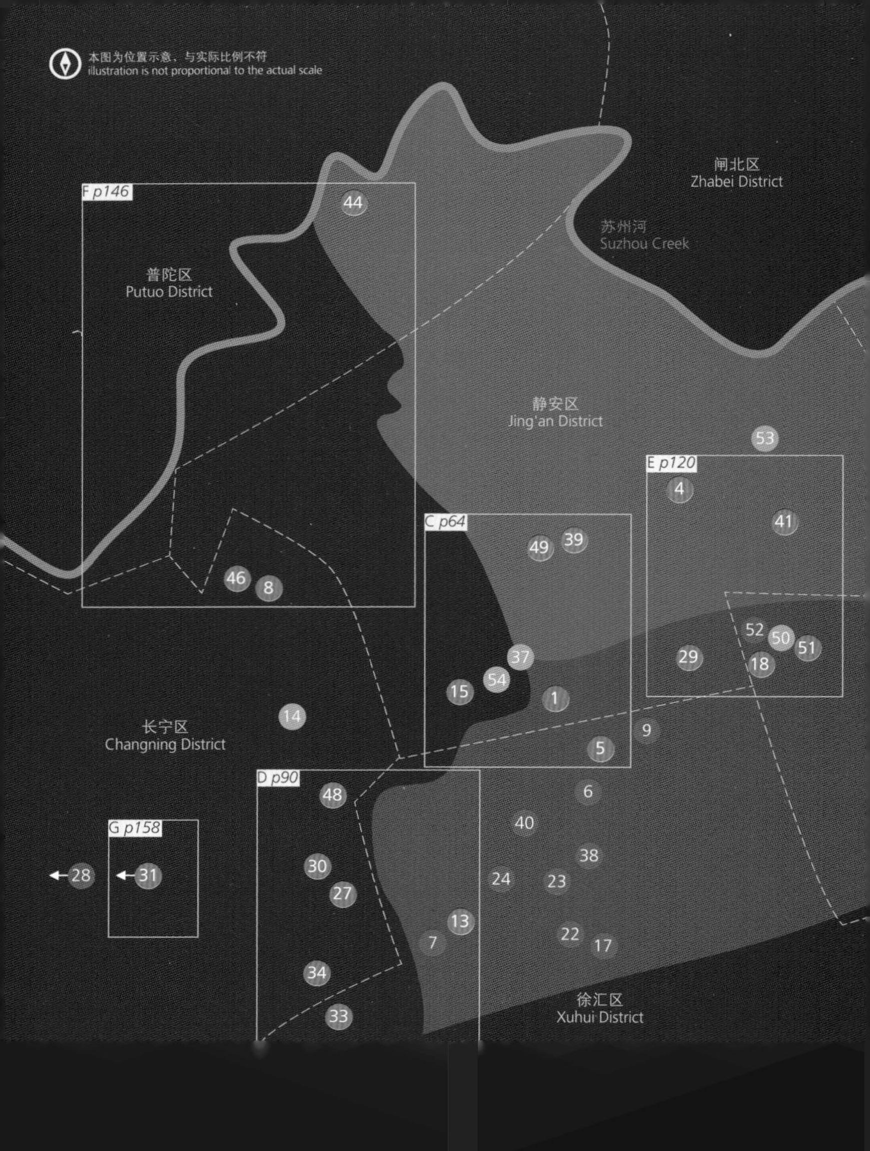
本图为位置示意，与实际比例不符
illustration is not proportional to the actual scale
闸北区
Zhabei District
苏州河
Suzhou Creek
普陀区
Putuo District
静安区
Jing'an District
长宁区
Changning District
徐汇区
Xuhui District
F p146
E p120
C p64
D p90
G p158
44
53
4
41
49
39
46
8
52
50
51
29
18
37
54
15
1
14
9
5
48
6
40
38
24
23
30
27
28
31
13
7
22
17
34
33

虹口区
Hongkou District
杨浦区
Yangpu District
47
25
B p48
35
36
3
2
20
11
A p28
19
16
45
10
42
32
26
12
黄浦江
Huangpu River
浦东新区
Pudong New Area
43
21
黄浦区
Huangpu District
图例 LEGENDS
原公共租界
Former International Settlement
原法租界
Former French Concession
原上海县城
Former Chinese City
超出地图范围
Beyond the Map Scope
A-G
分区编号
Number of Zones
p28
分区地图所在页码
Page of the Zone Map
n
现存+详细介绍
Existent with More Details
n
现存
Existent
n
已拆除
Demolished

A 区

32. 慕尔堂 (1926-1931)
今沐恩堂
西藏中路 316 号 (近汉口路)
黄浦区

42. 大光明大戏院 (1931-1933)
今大光明电影院
南京西路 216 号 (近黄河路)
黄浦区

45. 国际饭店 (1931-1934)
南京西路 170 号 (近黄河路)
黄浦区

Zone A

32. Moore Memorial Church
(1926-1931)
now Mu'en Church,
No. 316 Middle Xizang Rd
(near Hankou Rd), Huangpu District

42. Grand Theatre (1931-1933)
now Grand Cinema,
No. 216 West Nanjing Rd
(near Huanghe Rd), Huangpu District

45. Park Hotel (1931-1934)
No. 170 West Nanjing Rd
(near Huanghe Rd), Huangpu District

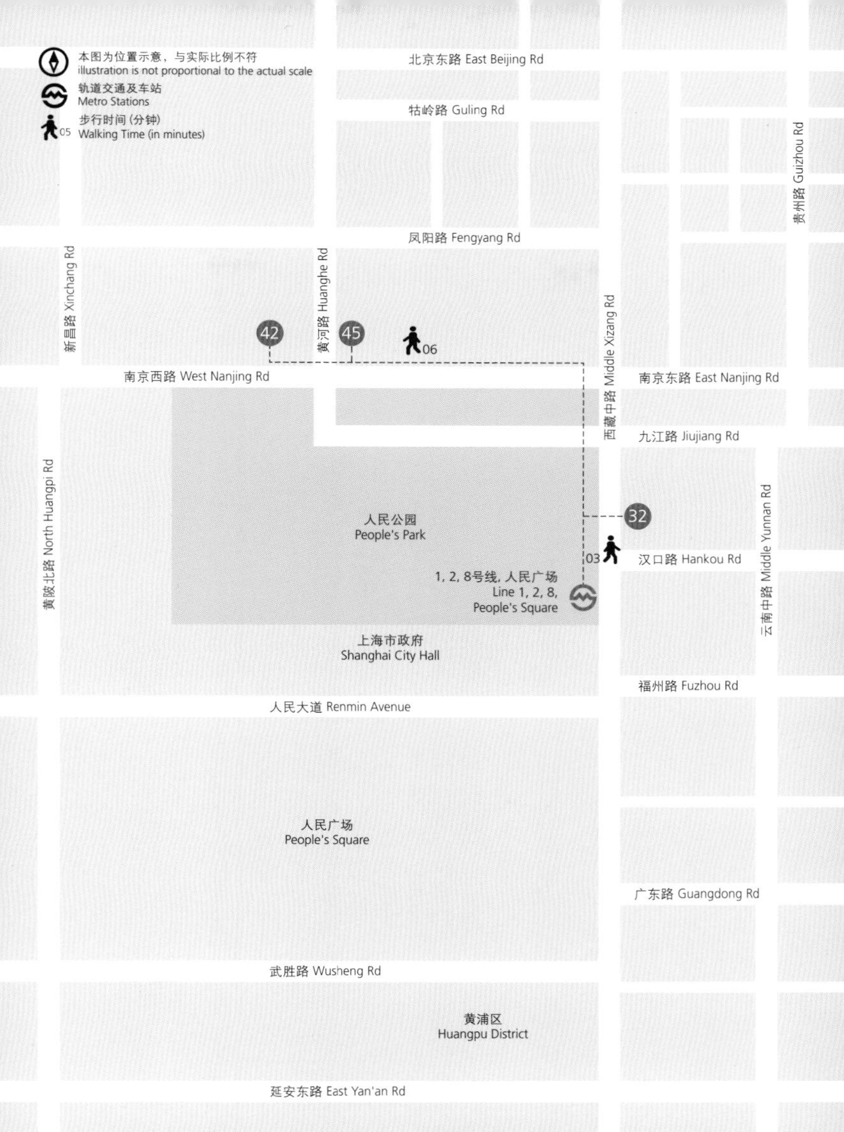

本图为位置示意，与实际比例不符
illustration is not proportional to the actual scale
轨道交通及车站
Metro Stations
步行时间 (分钟)
Walking Time (in minutes)
05
北京东路 East Beijing Rd
牯岭路 Guling Rd
凤阳路 Fengyang Rd
贵州路 Guizhou Rd
新昌路 Xinchang Rd
黄河路 Huanghe Rd
西藏中路 Middle Xizang Rd
42
45
06
南京西路 West Nanjing Rd
南京东路 East Nanjing Rd
九江路 Jiujiang Rd
黄陂北路 North Huangpi Rd
32
云南中路 Middle Yunnan Rd
人民公园
People's Park
03
汉口路 Hankou Rd
1, 2, 8号线, 人民广场
Line 1, 2, 8,
People's Square
上海市政府
Shanghai City Hall
福州路 Fuzhou Rd
人民大道 Renmin Avenue
人民广场
People's Square
广东路 Guangdong Rd
武胜路 Wusheng Rd
黄浦区
Huangpu District
延安东路 East Yan'an Rd

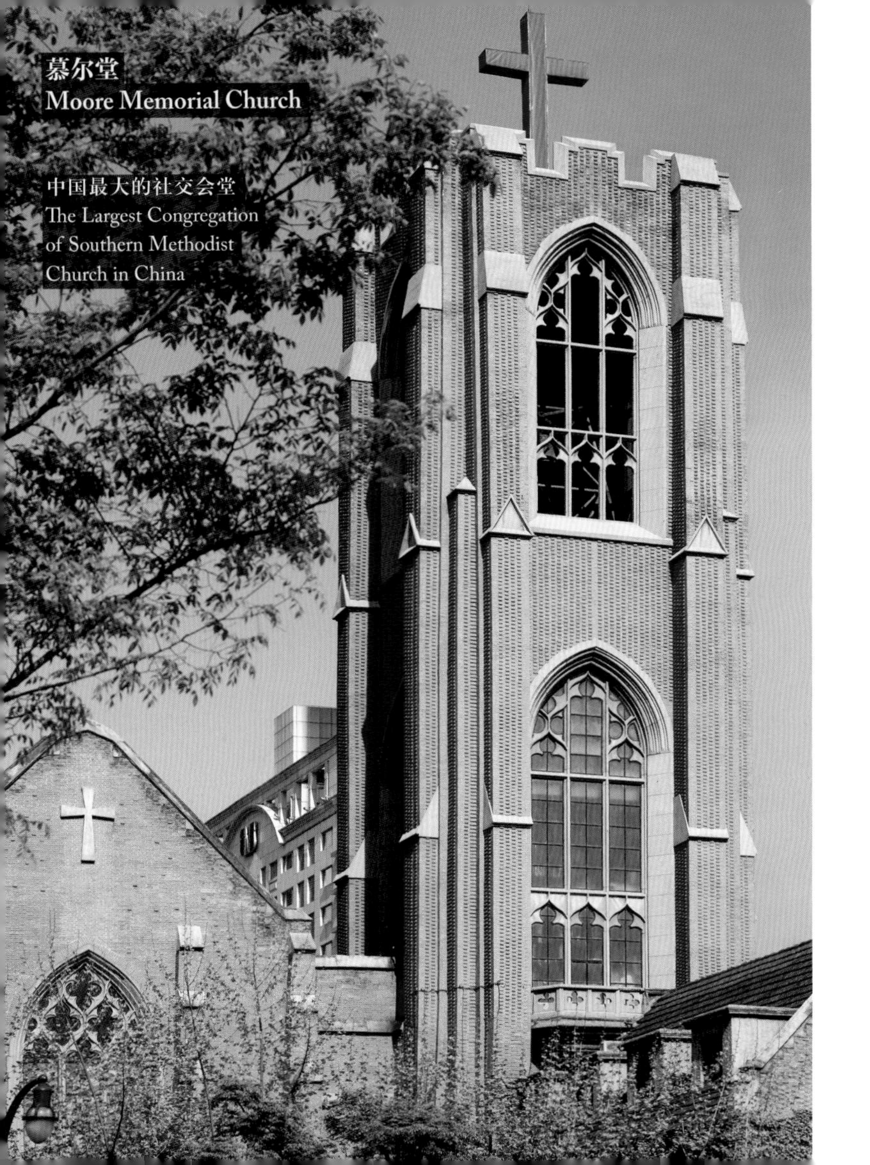

慕尔堂
Moore Memorial Church

中国最大的社交会堂
The Largest Congregation of Southern Methodist Church in China

慕尔堂始建于清光绪十三年（1887），原名监理会堂，由美国南方监理公会传教士李德创立，位于汉口路云南路（今扬子饭店位置）。三年后，美国信徒慕尔为纪念早夭的女儿捐资，故更名。

1917 年起，慕尔堂改为社交会堂，活动时间、内容、对象因此大大扩展，到 1925 年前后，教友总数已超过千人，旧堂不敷使用。众信徒通过募捐筹款，几经周折，1929 年起在西藏路汉口路原中西女塾校址上建造新堂。

虽然邬达克本人信奉路德教，但是作为当时上海耶稣会的首席建筑师，除建造原德国新福音教堂外，他还承接了其他教派的教堂设计。

作为当时中国最大的社交堂，慕尔堂不仅从事宗教活动，也具有救济、教育等功能。教堂当时每天开放，从婴儿到成人都有活动，还主办女校和夜校。因此，其平面由五个体量构成：中间是 1200 人会堂，其余四角分别是社会、教育、管理和娱乐部门，其中东北角为原来中西女塾旧楼。主入口设在朝向跑马厅的西藏路，在南北两个方向还分别设置了女教友入口、内院和操场。

两层高的大堂空间开阔，东端设唱诗班座席，木装饰精美。二层是三面围合的楼座，与西方古典歌剧院布局相似。大堂顶部作肋骨拱顶形式，外侧有扶壁。靠外廊两侧尖券窗饰以黄色玻璃彩绘，即使在阴天也像有微弱的阳光照射，营造出神秘的宗教气氛。

建筑外立面为美国学院哥特式风格，局部有罗马风手法。外墙饰以深红色面砖，且表面凹凸不平，形成独特的肌理。墙角及窗口镶有深灰色隅石。塔楼、大堂入口及大厅两侧均采用火焰式尖券窗。西南角钟楼高度居当时上海建筑之首，达 42.1 米。顶部安装着信徒捐赠的 5 米高霓虹灯十字架，底座还配有马达。

1931 年春，超过 10 个民族，1500 名中外教友参加了奉献（落成）典礼。

抗战期间，慕尔堂曾临时收容难民。1942 年后被日军侵占，大堂一度作为马厩，损毁严重。抗战胜利后，教会收回并修复教堂。1958 年，区内基督教各派实行联合礼拜后，改名沐恩堂。“文革”时，房屋被学校使用。1979 年 9 月恢复宗教活动。

1991 年，沐恩堂被列入上海市级文物保护单位。近年，因为周围兴建高层，教堂结构受损而大修，2010 年春已重新开放。

参观指南

周六周日开放。

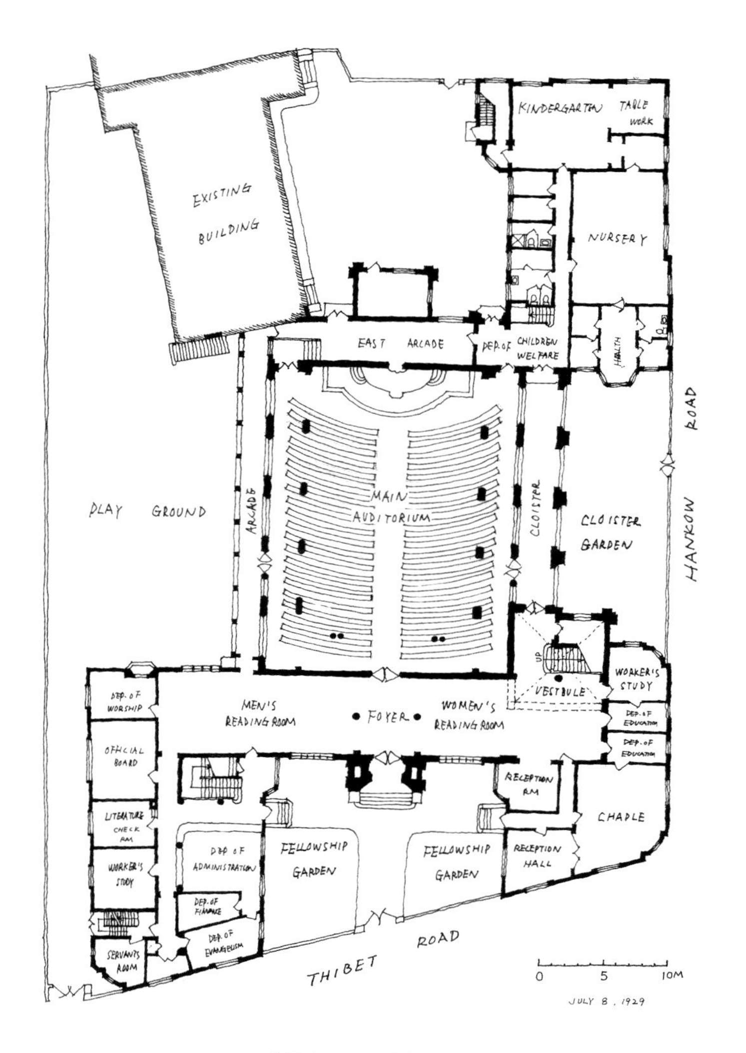

首层平面 Ground Floor Plan

The Moore Memorial Church was built in 1887 by American missionary C. F. Reid who belonged to the Southern Methodist Church. It was renamed Moore Memorial Church after American follower Mr. J.M. Moore donated to the church in 1890 in memory of his daughter.

The church also came to be used as a place for social and informal gatherings in 1917 by stretching its open time. By 1925, the number of its followers had increased to more than 1,200 — far too many for the church to handle. With more donations, the construction work for a new church finally kicked off in 1929 on the former site of the McTyeire School for Girls.

Hudec believed in Lutheran Church. As the chief architect of Shanghai Societas Jesus, he had designed the New German Lutheran Church as well as churches for other religious schools.

As China's largest church that involved itself in large-scale social activities, Moore Memorial Church also hosts charity and education programs.

The layout was essentially composed of five parts, including a main auditorium that could house 1,200 people in the center, and four other sections that dealt with issues relating to society, education, management and entertainment on the four corners. The main entrance was located on the Xizang Road facing the Race Course. There were two inner courts inside the church.

The two-story hall has an open space with exquisitely patterned wooden seats for choir in the eastern end. Enclosed on the three sides, the seats of the second floor are arranged in a layout similar to the western classic opera house. The Gothic vaulted ceilings finish with stone ribbings. The pointed arch window alongside the veranda is embellished with yellowish stained glass, which reflects vague sunlight even on a cloudy day and thus ensures a mysterious, religious atmosphere.

The façade is in a Collegiate Gothic style with Romanesque manners in some parts. The external walls are adorned with scarlet bricks in a textured pattern. The corners and windows of the church are graced by dark grey stone quoins. Flamboyant ecclesiastical lancet windows grace the tower. The bay windows are in the entrance and the flanking halls. The church has a 42.1-meter-high bell tower, which was one of the highest in the city at that time. The peak of the church is installed with a five-meter-high neon-light cross with motor on its base, which was donated by a follower.

The church had temporarily accepted refugees during China's Resistant War again the Japanese Aggression (1937-45). It was then occupied by the Japanese army, who used it as a horse stable and severely damaged the building. The church took it back after 1945 only to be used as a school during the Cultural Revolution (1966-76). Religious services have restarted since September 1979.

The church is listed as Shanghai Municipal Relics for Preservation since 1991. The new high-rise buildings in the neighborhood have harmed the architectural structure which underwent a big renovation in recent years and reopened to the public in the spring of 2010.

Tips

The church is open on Saturdays and Sundays.

大光明大戏院 Grand Theatre

远东最大的电影院

The Largest Cinema in the Far East

在邬达克的上海实践中，大光明大戏院的历史背景和基地条件都最复杂，但也最体现其设计功力。设计图纸最终为英国皇家建筑师学会收藏。

原址的老大光明影戏院于1928年闪亮面世，却因放映辱华电影《不怕死》声誉扫地，开业仅三年就被迫关门。1932年，英籍广东人卢根与美国商人创办联合电影公司，租赁大光明及其附近房产，用110万两白银，拆除旧屋，请邬达克设计重建，并更名为“大光明大戏院”。1933年6月14日大光明首映好莱坞电影《热血雄心》，开始其辉煌时代。

新中国成立前，大光明大戏院主要放映美国福克斯、米高梅等好莱坞的原版片，也是工部局音乐会的常驻地，还是第一家使用译意风（类似同声翻译）的影院，堪称引领摩登生活的时尚标杆。

大光明大戏院实际上是一座集影院、舞厅、咖啡馆、弹子房等于一身的娱乐综合体。这座曾经的“远东第一影院”生长在错综复杂的旧建筑夹缝中：沿静安寺路（今南京西路）门面不宽，大部分是需要保留的店面，沿百克路（今凤阳路）更是只有一条狭长的逃生通道。西面紧贴里弄，东侧与派克路（今黄河路）之间夹着卡尔登剧院（已拆除）——这座古典风格的影院是邬达克在克利洋行时期的作品。

因为基地狭长且不规则，设计几易其稿才最终确定，真正体现了“螺蛳壳里做道场”的功力。观众厅平行基地长边布置成钟形，与门厅轴线有30度扭转。大厅上下两层近2000个软座，容量当时居全国之首，内部采用暗槽灯照明，喷射式冷气。两层休息厅设计成腰果形，与流线形的门厅浑然一体。两部大楼梯直通二楼，休息厅中央还布置着灯光喷水池，噱头十足。

建筑外观是典型的现代装饰艺术风格，立面上横竖线条与体块交错。入口乳白色玻璃雨篷上方是大片金色玻璃，还有一个高达30.5米的方形半透明玻璃灯柱，夜晚尤为光彩夺目。

新中国成立后，大光明大戏院更名为大光明电影院，并历经数次改造，最后一次整修在尽可能恢复历史原貌的同时，更新了设施，新增了5个小厅、屋顶花园和餐厅，并于2009年1月19日重新开业。经历近80年风雨的大光明继续闪耀于上海顶级影院之列。

参观指南

每天10:00-23:00，建筑室内外均开放，放映厅需购票进入。西侧疏散通道内的“影院历史文化展廊”也颇具特色。

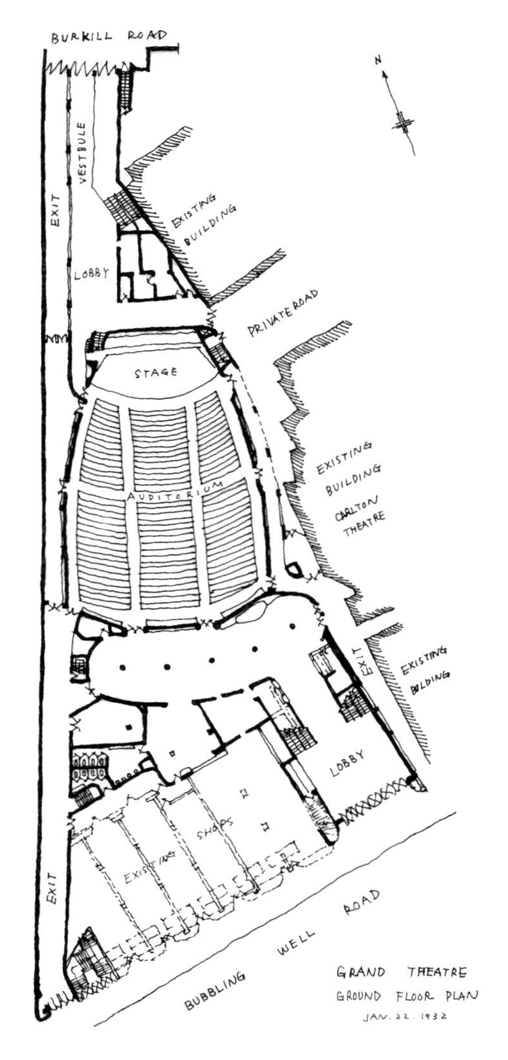

首层平面 Ground Floor Plan

The Grand Theatre was the most difficult job for Hudec among his major achievements in Shanghai. He showed the ultimate skills in handling this work which almost looked like mission impossible — to design a spacious, stylish cinema on a strange-shaped base. The draft was later added to the collection of the Royal Institute of British Architects.

On the same site there used to be an old cinema, which earned notoriety after screening "Welcome Danger," a movie that humiliated the Chinese. It was forced to close down in 1931 after opening for only three years.

British Cantonese Lu Geng founded the United Movies Co. with American merchants in 1932. He rented the site, demolished the old cinema and invited Hudec to design a brand new one. "Grand Theatre" opened its doors to the public on June 14, 1933, with the Hollywood movie "Hell Below". It became a popular joint since then, going on to screen movies produced by 20th Century Fox and Metro-Goldwyn-Mayer, and staging concerts of the Municipal Council. It was the first cinema to offer simultaneous translation ("Earphone") during the screening of foreign films.

Grand Theatre was a complex work of cinema, dancing hall, cafe and billiard rooms. The cinema perched on a long, narrow and irregular base, almost on the cracks of the surrounding old buildings. Hudec's talent in utilizing spaces was fully showcased.

The auditorium was shaped like a big bell. It could seat about 2,000 people on two floors — the biggest capacity in any Chinese movie halls. The interior was illuminated by cove lightings and cooled by jet-refrigerator air supply scheme.

Graced by fancy lighted fountains, the lobbies on the first and second floors were shaped like cashews to fit well with the base. Two grand, stylish staircases led audience from the entrance straight up to the second floor.

Grand Theatre was a typical Art Deco building. The façade was embellished with flowing lines and vivid colors. A large piece of golden-hued glass topped the ivory glass canopy at the entrance. Together with a cubic translucent glass lighting pillar up to 30.5 meters high, they glistened beautifully at night.

Grand Theatre was renamed Grand Cinema after 1949 and endured numerous renovations since then.

A recent renovation not only revived its original look, but also upgraded the facilities and added five small cinema halls, a roof garden and a restaurant.

On January 19, 2009 the cinema reopened after 80 years and the Grand Theatre was once again glistening as one of the city's top cinematic theatres.

Tips

10am to 11pm, open to the public. Tickets are required to enter the auditorium. An exhibition showcasing history of cinema at western exit gallery is worth a visit.

国际饭店 Park Hotel

远东第一高楼
The Tallest Building in the Far East

上海四行储蓄会二十二层大厦（地上 22 层，地下 2 层，总高 83.8 米），即国际饭店，无疑是邬达克建筑作品中质量最高、影响最大的一个。这座大厦曾多年享有“远东第一高楼”的美誉，并保持上海建筑高度神话长达半个世纪，楼顶旗杆的中心位置还被定义为上海城市测绘的零坐标。

1930 年，由金城、盐业、大陆和中南四家银行联合组成的四行储蓄会已通过“储蓄分红”快速积累了大量资金，看到地价飞涨、房地产业利润丰厚时，决定投资高层现代旅馆，选址在面对跑马厅的静安寺路派克路转角，建筑英文名“派克饭店”正由此而来。

邬达克能最终赢得设计竞赛，不仅因为两年前四行储蓄会汉口路联合大楼的成功在业主那里积攒了人气，或是一年前在纽约、芝加哥等地观摩美国摩天楼带来了设计灵感，更是因为在高层建筑结构和技术上大胆突破，最主要是解决了上海软土地基处理的致命问题。400 根 33 米长的木桩和钢筋混凝土筏式基础，上层采用质量轻、强度大的合金钢结构，这些举措使国际饭店在同时期兴建的上海高层建筑中沉降量最小。

大楼底层主要是四行储蓄会的营业大厅，金库设在地下室，转角才是旅馆门厅，这跟今天的状况正好相反。二层餐厅朝南是大面积出挑的落地玻璃窗，可以一览无遗地俯瞰跑马厅。

建筑立面强调垂直线条，层层收进直达顶端，表现出美国现代派装饰艺术风格的典型特征。高耸且稳定的外部轮廓，尤其是 15 层以上呈阶梯状的塔楼，在四周早已高楼林立的今天仍显得雅致动人。

国际饭店综合了世界各国的先进技术，更见证了中国近代施工行业的奇迹。当时刚完成南京中山陵的馥记营造厂主导了施工。建筑外墙基座采用山东产黑色抛光花岗石，上部则是深褐色的泰山面砖。

1934 年 12 月 1 日，国际饭店由时任上海市市长吴铁城启幕，中外媒体广泛报道。作为现代生活和国际都市的标志，当时的社会精英都以在此入住或举办社交活动为荣。它也是国际贵宾访沪的主要下榻处，“仰观落帽”的说法则在民间广为流传。美籍华裔著名建筑师贝聿铭也声称，正是被国际饭店的设计打动而选择了建筑学专业。

新中国成立后，特别是 1980 年后，国际饭店历经多次重新装修，现为锦江集团旗下的四星级酒店。2006 年被列为全国重点文物保护单位。

参观指南

建筑室外及室内公共部分可供参观，客房部分不对外开放。加拿大学者 2001 年带来的邬达克档案馆的珍贵历史图片，至今仍在大堂夹层的走廊内展出。

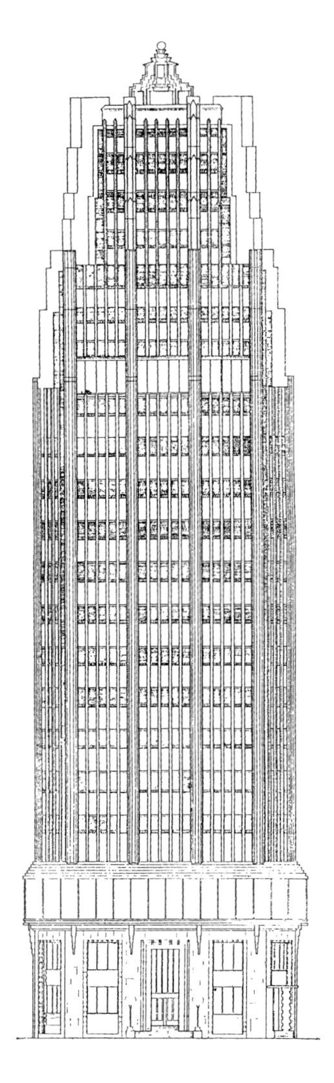

正立面 Front Elevation

The most famous work of Hudec must be the Park Hotel, a 22-story building designed for the Joint Savings Society. At a height of 83.8 meters with 22 floors and two underground basements, the once tallest building in the Far East had dominated the city's skyline for nearly half a century. Located in what is referred to as the "Zero Center Point of Shanghai'," the hotel is also the focus of the city.

In 1930, the Joint Savings Society, founded by four Chinese banks, had accumulated a quantity of capital through "active interests on deposits." Attracted by the soaring land price and handsome profits in real estate business, the society decided to invest in building a tall modern hotel at the corner of Bubbling Well Road and Park Road fronting the Race Course. The name "Park Hotel" originated from the location.

Hudec's success in winning the project from an open competition was largely due to his previous work on the Union Building for the society on Hankou Road.

He had also seen and was inspired by the skyscrapers in New York and Chicago during a trip to the US just a year before. Most importantly, he made a technological breakthrough in building a colossal building on the alluvial soil of mud and sand, a problem that had puzzled architects in Shanghai for many years.

The hotel sits on a reinforced concrete raft base of 400 33-meter-long piles of Oregon pine, which is topped by light-weight alloy with great strength. The structure caused the least sinkage among high-rise buildings in Shanghai at that time.

The ground floor used to be the banking hall of the society, which had a treasury in the basement. The entrance hall was on the corner of roads which was different from today. The restaurant facing south on the second floor featured large, protruding French windows that offered a great view of the Race Course.

The façade is emphasized on vertical stripes which are shrinking layer upon layer until the top, a typical American modern Art Deco style. The imposing but stable silhouette, as well as the staircase-like tower above the 15th floor, gives a unique elegant look among surrounding modern skyscrapers in the concrete jungle today.

The construction of the Park Hotel had absorbed advanced technologies from all countries including the incredible quality of Chinese modern constructions. The main contractor Voh Kee Co. was one of the biggest Chinese companies in the field of constructions, which had just completed the project of Dr. Sun Yat-sen's Mausoleum in Nanjing.

The hotel is also made of Chinese materials. The plinth of the external walls is composed of black polished granite from Shandong Province in north China. The facades are faced with dark brown Taishan facing tiles.

Shanghai mayor Wu Tiecheng attended the opening ceremony of the Park Hotel on December 1, 1934, which was widely reported in the Chinese and foreign media.

As an icon venue of modern life, it was an honor for social elites to stay or host a social event in the Park Hotel at that time. The hotel was also a first choice for international VIPs during their Shanghai trips. Worldly renowned American Chinese architect I. M. Pei has admitted it was the Park Hotel that had greatly aroused his interest to study architecture, which was against the will of his banker father.

The Park Hotel has endured several renovations since 1949 and now serves as a four-star hotel of the Jinjiang Group. It was listed as a national heritage building for preservation in 2006.

Tips

The hotel is open to the public but the hotel rooms are private. Precious historical pictures taken by a Canadian scholar in 2001 from the Hudec Archives are exhibited on the corridor of the mezzanine.

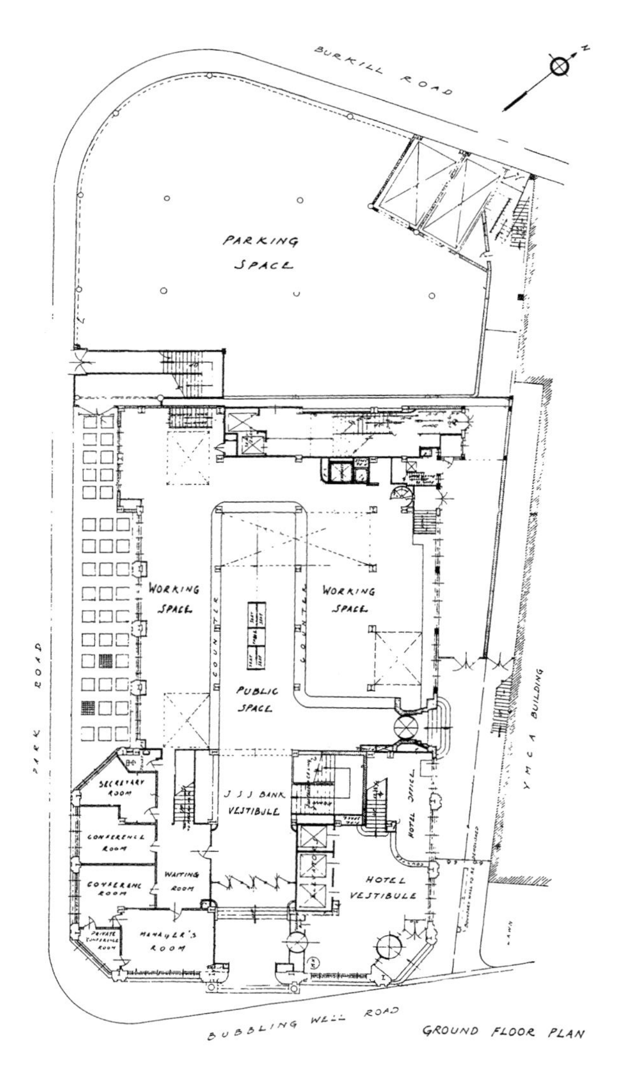

首层平面 Ground Floor Plan

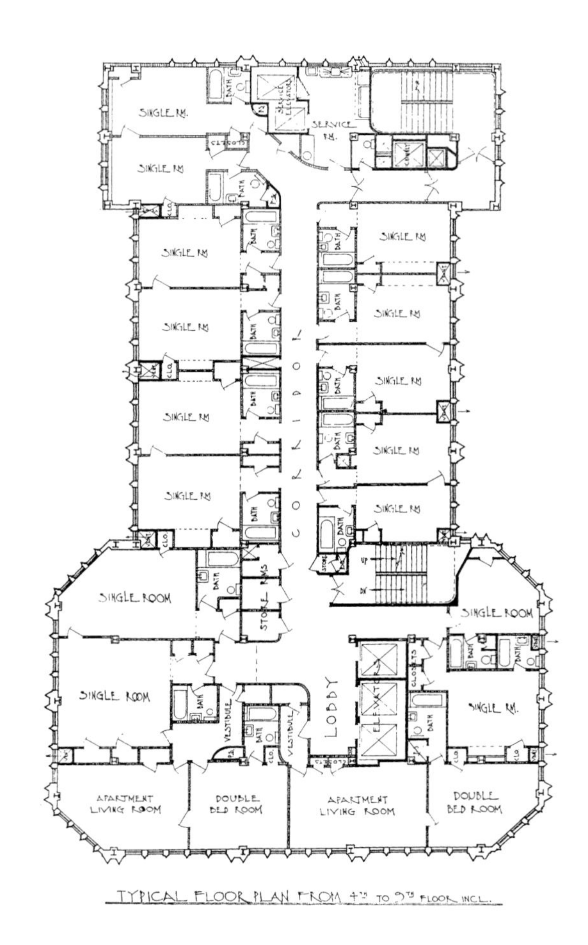

标准层平面 Typical Floor Plan

B 区

11. 美国总会（1922-1924）
    原高法大楼
    福州路 209 号（近河南中路）
    黄浦区

20. 四行储蓄会联合大楼（1926-1928）
    今联合大楼
    四川中路 261 号（近汉口路）
    黄浦区

35. 广学大楼（1930-1932）
    虎丘路 128 号（近香港路）
    黄浦区

36. 真光大楼（1930-1932）
    圆明园路 209 号（近香港路）
    黄浦区

Zone B

11. American Club (1922-1924)
    former Shanghai High People's Court
    No. 209 Fuzhou Rd (near Middle He'nan Rd), Huangpu District

20. Union Building of the Joint Savings Society (1926-1928)
    now Union Building
    No. 261 Middle Sichuan Rd (near Hankou Rd), Huangpu District

35. Christian Literature Society Building (1930-1932)
    No. 128 Huqiu Rd (near Xianggang Rd), Huangpu District

36. China Baptist Publication Building (1930-1932)
    No. 209 Yuanmingyuan Rd
    (near Xianggang Rd), Huangpu District

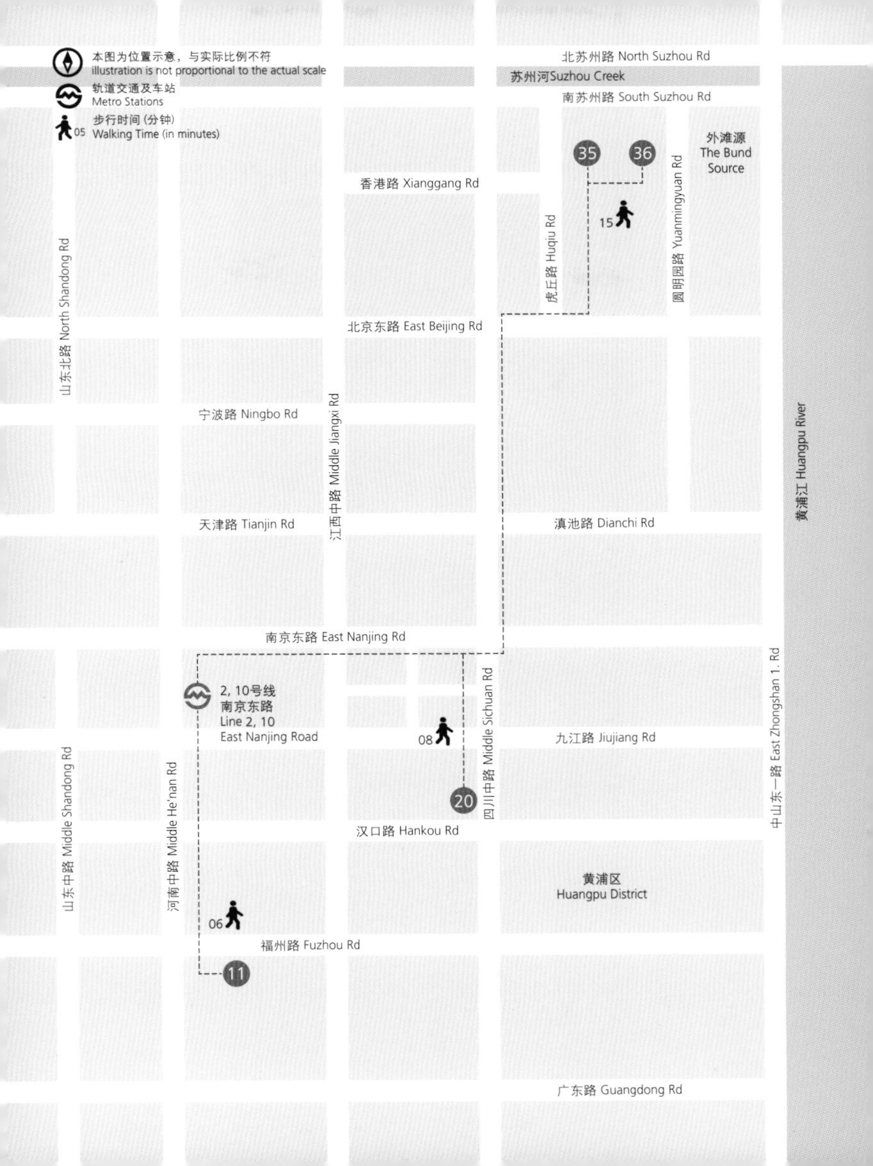

本图为位置示意，与实际比例不符
illustration is not proportional to the actual scale

轨道交通及车站
Metro Stations

步行时间（分钟）
05 Walking Time (in minutes)

北苏州路 North Suzhou Rd

苏州河Suzhou Creek

南苏州路 South Suzhou Rd

外滩源
The Bund
Source

35

36

15

香港路 Xianggang Rd

虎丘路 Huqiu Rd

圆明园路 Yuanmingyuan Rd

山东北路 North Shandong Rd

北京东路 East Beijing Rd

宁波路 Ningbo Rd

江西中路 Middle Jiangxi Rd

天津路 Tianjin Rd

滇池路 Dianchi Rd

黄浦江 Huangpu River

南京东路 East Nanjing Rd

2, 10号线
南京东路
Line 2, 10
East Nanjing Road

08

四川中路 Middle Sichuan Rd

九江路 Jiujiang Rd

中山东一路 East Zhongshan 1. Rd

山东中路 Middle Shandong Rd

河南中路 Middle He'nan Rd

20

汉口路 Hankou Rd

黄浦区
Huangpu District

06

福州路 Fuzhou Rd

11

广东路 Guangdong Rd

美国总会 American Club

旧上海三大外侨总会之一
One of the Top Three Clubs for Expatriates Living in Shanghai after World War I

美国总会与上海总会（即英国总会）、法国总会合称“一战”后旧上海三大外侨总会。其前身是1916年一群美国桥牌爱好者建立的“大满贯俱乐部”，1917年5月宣布成立“美国人俱乐部”，即“美国总会”。俱乐部采用会员制，正式会员必须是居住在上海的男士，入会费用100美元。其他国家人士也可加入，但人数有限。1945年前俱乐部不接受女会员。女士每年仅有一次机会参加活动，那一晚称为“女士之夜”。

1922年初，总会盘下福州路地皮筹建新总部。同年8月，克利洋行在16家英美建筑事务所参与的设计竞赛中胜出。新大楼于1923年5月破土动工，1924年11月正式落成。

美国总会是邬达克在克利洋行时期的代表作，占地916平方米，总建筑面积6 753平方米，高7层（包括半地下室一层），全钢筋混凝土结构。内部设施完备先进：地下室有保龄球房、健身房、土耳其浴室；首层是大型桌球房、酒吧；二层设麻将室、棋牌室、阅览室；三至五楼包含50个标准客房；六层设会议厅、大宴会厅；屋顶花园可眺望黄浦江景观。整座大楼配备暖气，贴着主楼梯还有2部客梯。

建筑外观呈美洲殖民地时期乔治式风格，这也是当时美国本土俱乐部普遍采用的形式。立面对称布局，竖向分三段。入口为三开间的浅门廊，采用白色大理石塔司干柱式；二层设落地长窗，外加铸铁栏杆；顶层为白色大理石的双壁柱圆拱券窗。外墙主材为美国进口的棕色面砖，窗上有白色平券状楣饰。

室内也是美洲殖民地风格：深色木装修，有古典浅壁柱及白色门框，地面用米色大理石或深色橡木地板。门厅内对称的大理石弧形楼梯有精美的铸铁花饰。最鲜为人知的是，二楼麻将室由中国专家设计，采用正宗官式风格，顶部有藻井和彩画，悬挂宫灯，家具配四仙桌和太师椅。

新中国成立后，该大楼成为上海市高级人民法院和中级人民法院所在地，俗称“高法大楼”。1991年，高、中级法院迁至虹桥路。2005年后大楼撤空等待新的开发利用。

2008年6月，中国第三个文化遗产日之际，在该楼顶层大厅里，上海市城市规划管理局和匈牙利驻上海总领事馆合作举办了“建筑华彩——邬达克在上海”的展览。

参观指南

福州路可欣赏建筑的外立面，室内不开放。

The edifice on 209 Fuzhou Road was a new version of the premier all-male American club when it was completed in November 1924, a masterpiece by Hudec during his time at R.A. Curry's firm.

Founded in 1917 at the site of the former "Slam Club," American Club was one of the top three clubs for expatriates living in Shanghai after World War I. The other two were Shanghai Club (mainly for the British) at No. 2 on the

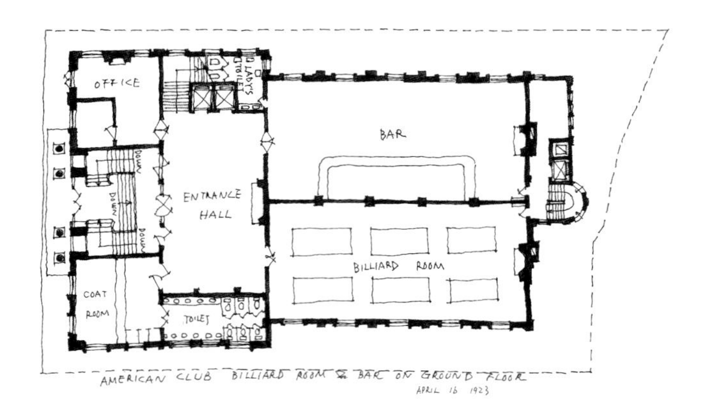

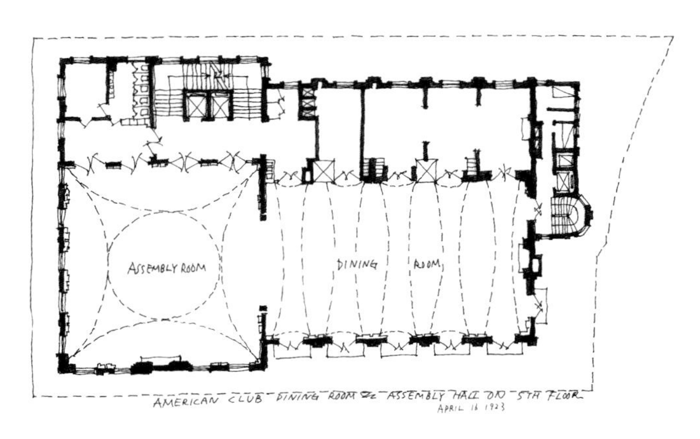

首层、六层平面 Ground Floor Plan, 5th Floor Plan

Bund and French Club (first on Yandang Road, later moved to the building which is the Garden Hotel today).

The club was open to members only — men living in Shanghai. The memberships cost US$100 per person and were limited to other nationalities. Before 1945, women members were not allowed. However, they did get a chance once a year for "the LADIES Night."

In 1922, the American Club bought the site on Fuzhou Road. R.A. Curry's firm won the rights to build the club at an open competition which included 16 British and American architecture firms. The construction started in May 1923.

Covering 916 square meters, the seven-story building (including the basement) was built with a steel and concrete structure and covered a building area of 6,753 square meters.

Equipped with advanced facilities, the building had almost everything that a gentleman needed to enjoy, including a bowling room, a gym and a Turkish bath in the basement; a bar and a grand billiard space on the ground floor; a mahjong room, a card room and a reading room on the second. From third to the fifth floor, there were 50 bedrooms. An assembly room and a banquet room were on the sixth floor while the roof garden overlooked the Huangpu River. The whole building was equipped with heating system and boasted two elevators.

For the American Club, Hudec chose American -Georgian colonial style architecture, a popular style in vogue for clubs in the U.S. at that time. The symmetrical elevation is in a classic three-section vertically. A three-bay porch with white marble Tuscan orders graced the entrance. The second floor was embellished with French windows and iron railings. The top floor featured double-pilaster arch windows. The external walls were covered by US-made brown facing tiles, which were highlighted by white flat arch window lintels above the windows.

The interior had an American colonial look covered with lots of dark wood, classic applied columns and white door frames. The floor was paved with either beige-hued marble or dark-toned oak wood. The pair of symmetrical curved marble staircases were adorned with exquisite cast iron patterns. Very few know that the mahjong room was done by a Chinese expert that ensured local flavor. The room had a Chinese-style coffered ceiling with lots of colorful paintings and palace lanterns decorated as ornaments. It was also furnished with traditional Chinese square tables and fauteuils to enhance its Chinese ambience.

After 1949, the American Club became the offices for the Shanghai High People's Court and Shanghai Second Intermediate People's Court. After the two courts moved to Hongqiao Road in 1991, the building has been lying vacant since then.

In June 2008, an exhibition of Hudec's heritage works in Shanghai was co-hosted by Shanghai Urban Planning Bureau and the Hungarian Consulate General on the top floor of the former American Club. The building remains one of Hudec's early masterpieces in the city.

Tips

The building is not open to the public but you can appreciate the façade from Fuzhou Road.

四行储蓄会联合大楼
Union Building of the Joint Savings Society

沪上首座外立面大量用大理石厚板的建筑
The City's First Building Whose Façade was Lavishly Adorned by Heavy Marbles

四行储蓄会成立于 1923 年 6 月，是由金城、盐业、大陆、中南四家银行组成的联营事务所（旧中国银行业中第一家，也是唯一一家联营组织）共同创建的经济实体。

1926 年，四行储蓄会在四川路汉口路转角筹建新总部大楼，亦称“联合大楼”。这是邬达克首次为中国业主设计大型公共建筑，8 年后，这对出色的甲方乙方组合还将共同谱就“远东第一高楼”的神话。

联合大楼位于道路转角，但基地偏薄。当时法规要求建筑转角必须呈圆形，这样一来大楼会完全失去体量感。邬达克对此非常不满，曾嘲讽道：“自然界唯一圆的东西是老头的肚子，然而上帝禁止我们以此为原型设计建筑细部。”为了既不违反法规，又能利用好基地，他在道路转角的圆弧上设置了巨大的拱券作为银行主入口，入口上方平面则改为八边形，确定的明暗关系增强了建筑的体积感。八边形塔楼直接升起成为制高点，显得简洁有力。

大楼半地下室是金库及兑钞室，一层是饰有爱奥尼柱式的门厅和饰有陶立克柱式的营业大厅，地面铺淡黄色大理石，台面以及精美的装饰均由白色大理石雕刻而成，上有铜制矮栏，出纳窗口则被雕琢成西方古典门洞式样。二层是银行主要办公区，三至七层为出租办公区，八层转角设餐厅。

大楼外形呈英国乔治式折中主义风格。立面竖向分三段。基座、顶层以及窗楣为白色大理石，三至六层主体为深褐色的泰山面砖。沿四川路四根高大的科林斯式圆柱雕刻精美，柱间通高的落地长窗为营业大厅带来充足的光线。室内外还有大量精美的卷叶饰。

邬达克曾为这座大楼的整体布局和细部装饰耗费过许多不眠之夜，更首次在外立面上大量使用从日本进口的白色大理石厚板，并以意大利工艺雕琢成柱式或雕带，这在当时通常采用花岗岩饰面的上海颇为罕见。室内材料则呈现丰富的颜色和质感。设计者认为，建筑不应过度装饰，但平衡颜色和材料的关系至关重要，也明确表示偏爱丰富多彩而非灰色的建筑。

联合大楼的塔楼据说是邬达克对匈牙利北部乡村式文艺复兴风格的追忆。建筑师自己设计的大楼自动铜制卷帘门由布达佩斯的一家金属加工厂生产。

新中国成立后，联合大楼归上海化工轻工供应公司所有，大厅部分后为广东发展银行营业大厅。大楼现正等待新的开发利用。

参观指南

四川中路与汉口路可欣赏建筑的外立面，室内不开放。

The Joint Savings Society, which was formed in June 1923, was a co-operative venture between the four big northern Chinese banks — Kincheng, Yien Yieh, Continental and the China and South Seas banks. It went on to become a successful venture and an influential Chinese financial institution of the time.

When the society planned to build a new office, the "Union Building" in 1926, Hudec won the rights for the project from an open competition. It was his first big project for a Chinese owner. And only eight years later the

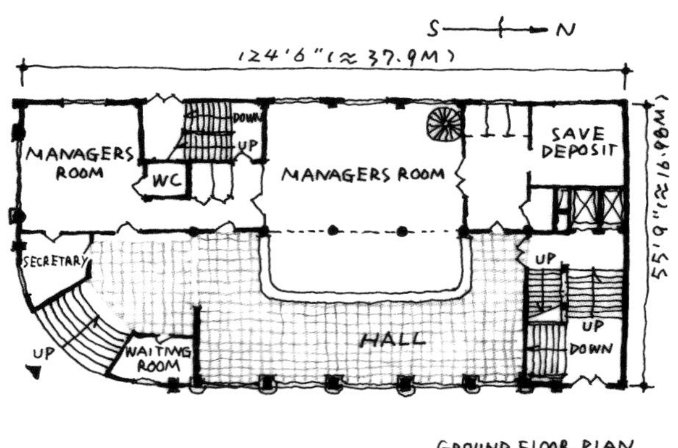

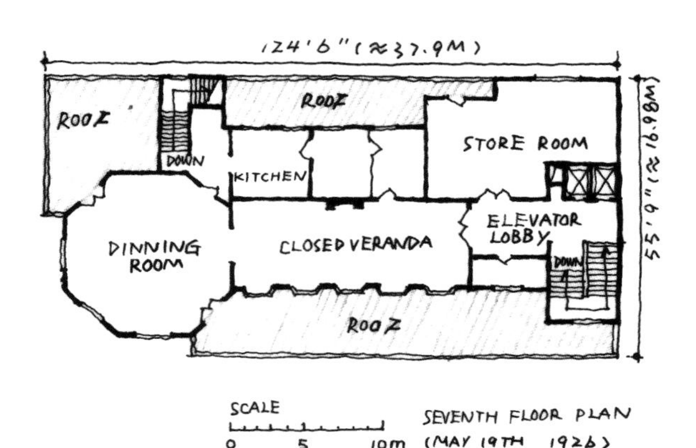

首层、八层平面 Ground Floor Plan, 7th Floor Plan

society and the architect came together again to build the landmark Park Hotel, the once tallest buildings in the Far East.

When the "Union Building" threw open the doors, the English newspaper "Far Eastern Review" said it was "one of the most unique buildings in the business section of the town."

It was built in an eclectic British Georgian style. The elevation was in three sections vertically.

The main entrance was located at the intersection of the two roads. It was an immense marble archway which was closed by a heavy bronze chain curtain. An octagon tower, which was believed to be Hudec's memory of the Renaissance style in northern Hungary, soared to the top of the building, making for a simple and powerful effect.

The rest of the façade, from the third to the sixth floors, was made with deep brown Taishan tapestry bricks, marking a pleasant contrast with the shining white marble of the lower stories.

Although very simple in design, the building impressed essentially due to the beauty of the true materials that were selected for facing the elevation.

The elevation was embellished with white Carrare marble showing off the peculiar classical touch in the treatment of the columns, cornices and ornamentations.

French windows between the columns introduced ample light into the banking hall. Inside and out, the building was graced by plenty of delicate ornamentations of Acanthus leaf.

In many sleepless nights, Hudec had crushed his brains to perfect the layout and the architectural details. It was his innovation to lavishly use white marbles on the façade with carved Italian columns and belts, which was rare among the usual granite façade in the city.

The interior materials showcased a rich layer of colors and textures. Hudec believed it was crucial to balance colors and materials. However he preferred rich-colored buildings than grey ones.

The building was very functional. The joint treasury was located in the semi-basement, which was attached to special money cashing rooms and a sycee vault equipped with elevators and a sycee box transporting system.

The main banking hall on the elevated ground floor was treated with marble wall facings, marble counters and columns in delicate tones and subdued simple designs. The society occupied the second floor while the third to the seventh floor were rented out as offices. The topmost floor featured the clubrooms for the employees of the Joint Savings Society and Treasury. The club consisted of a large lounge facing the roof terrace and pergola and various dining rooms.

The Union Building was used by a state-owned chemical company after 1949, part of which was later used as the banking hall for the Guangdong Development Bank. The building is not being used now.

Tips

The façade can be appreciated from Middle Sichuan Road and Hankou Road. The building is not open to the public.

广学大楼
Christian Literature Society Building

真光大楼
China Baptist Publication Building

装饰艺术派和表现主义建筑代表作，
邬达克洋行所在地

A Signature Building of Art Deco and Expressionism Which Housed the Hudec Firm

广学会是一个跨教派的基督教出版机构。其前身是1887年苏格兰传教士韦廉臣在上海创办的同文书会，1892年改为现名，含义是“以西国之新学广中国之旧学”。旗下主要出版物《万国公报》是当时中国发行量最大，也是最有影响的杂志，1899年曾最早向国人引介马克思及其《资本论》。

真光大楼即中华浸会书局大楼或浸信会大楼，机构前身是美国浸信会传教士湛罗弼于1899年在广州成立的美华浸会书局，旗下有中国基督教最早创办的月刊——《真光》(1902)，刊名取自《约翰福音》:“那光是真光，照亮一切生在世上的人。”

广学大楼与真光大楼是连体姐妹楼，其中广学大楼朝西，高九层，地下一层，真光大楼朝东，高八层，两者合成“U”形平面。主入口、电梯和卫生间分别设置，消防楼梯、锅炉房、水泵房合用。

广学会主要空间在底层和顶部几层。沿街设置开架书店，七层为员工宿舍，八层办公，九层为董事室和大阅览室，中间各层则出租营利。

真光大楼底层为浸信会的出版、印刷部门和零售商店，二、七两层为浸信会在沪组织办公处和学校，其中包括沪江大学第一任华人校长刘湛恩创办的城中区商学院。从大楼建成到1947年，整个八楼一直是邬达克洋行的办公地。

大楼整体为现代派装饰艺术风格，对哥特式尖券和壁柱的简化处理则体现了德国表现主义建筑的影响。一层外墙为石材贴面，二层以上为深棕色泰山砖，白色砂浆勾缝。其中最有特色的是窗间壁柱，事实上是对哥特式束柱的简化，强调垂直向上感。

邬达克这一时期作品中对砖砌工艺的娴熟运用，很可能也是他的弟弟盖佐的功劳，后者在美国学习半年后，于1930年6月来上海协助兄长，直到1933年2月因病在沪去世。

新中国成立后，广学大楼改为上海文体用品进出口公司，浸会书局移居香港，真光大楼也更换了主人。2002年起，该区域被纳入黄浦江两岸开发工程的“外滩源”项目一期范围。

参观指南

沿虎丘路、香港路、圆明园路可欣赏建筑的外立面，室内不对外开放。

The Christian Literature Society Building and the China Baptist Publication Building is a pair of twin buildings standing on the Yuanmingyuan Road. Work on both the buildings started in 1930 and was completed in 1932.

The Christian Literature Society was an influential publishing organization of its time, which had developed on the foundation of the Society for the Diffusion of Christian Knowledge among the Chinese founded by American Presbyterian missionary Alexander Williamson, who was from Scotland. The society was among the first to introduce western publications to China including Karl Marx's classic "Capital".

The China Baptist Publication Society Building was formerly known as Gospel Light Building, for the publications of Gospel Light by the Baptist organization founded by American missionary Robert E. Chambers.

The U-shaped twin buildings were linked with each other and had a similar layout.

The nine-floor Christian Literature Society Building faced west while the eight-floor China Baptist Publication Building looked east. They had their own entrances, elevators and washing rooms but shared a fire staircase, a boiler house and a pump house.

The Christian Literature Building featured a book store on the ground floor, dormitories for employees on the seventh floor, office on the eighth and boardroom as well as a grand reading room on the top floor. The remaining floors were leased out.

The Baptist Building contained publishing department and a retail shop on the first floor and offices and some important organizations from the second to the seventh floors, such as the Christian Publishing House and Downtown College of Commerce. Hudec's firm occupied the entire eighth floor from 1932 to 1947.

Both the buildings were designed in Art Deco style. The dark brown tiles on the façade, the horizontal lines and the setback structure over the top showcased typical Art Deco elements, which were popular among American and Shanghai's skyscrapers at that time.

The façades were decorated by acute-angel shaped lines, rolled upon the parapets. Hudec's simplified treatment of Gothic pointed arches and pilasters revealed the influence from German expressionism architecture.

The twin buildings captured the maturity of Hudec's own architectural style, which had taken the lead in the Gothic Revival architectural features, using tiles as wall-coating that reflected a new style.

In addition, his masterful use of brick laying during the period probably attributed to his brother Geza, who had studied in the US for half a year and came to Shanghai to assist his brother in June of 1930. Unfortunately Geza died of illness in February 1933.

After 1949, the Christian Building became the Shanghai Stationery and Sports Goods Import and Export Company while the Baptist Building also changed owners after China Baptist Publication moved to Hong Kong.

Since 2002, the twin buildings were included in the Waitanyuan project of the Huangpu District.

Tips

The buildings are not open to the public. However you can appreciate their facades along Huqiu Road, Xianggang Road and Yuanmingyuan Road.

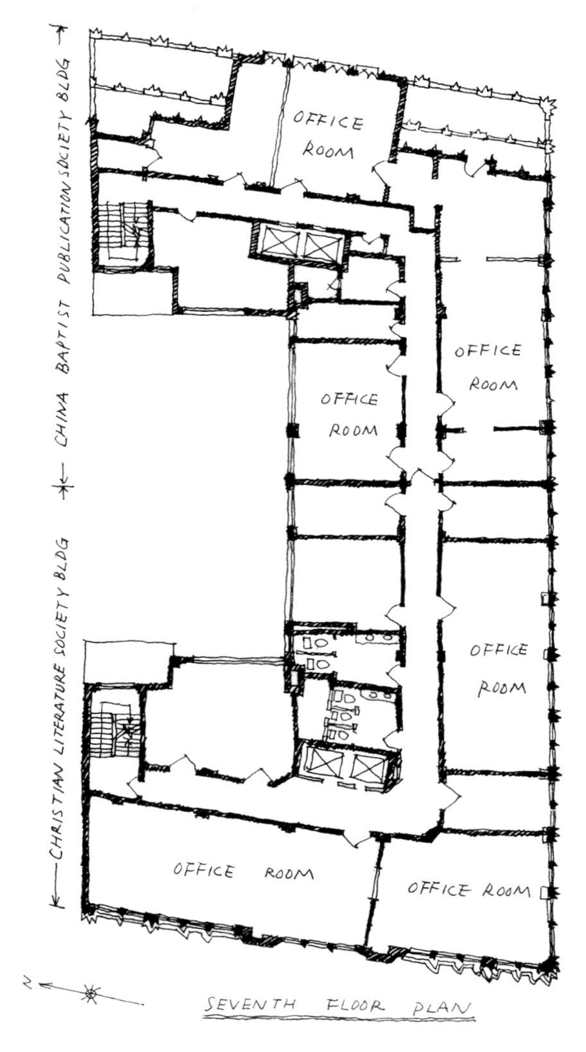

八层平面 7th Floor Plan

C 区

1. 巨籁达路 22 栋住宅 (1919-1920)
   巨鹿路 852 弄 1-10 号、
   巨鹿路 868-892 号（近常熟路）
   静安区

5. 盘滕住宅 (1919-1920)
   今仙炙轩
   汾阳路 150 号（近桃江路）
   徐汇区

15. 宏恩医院 (1923-1926)
    今华东医院 1 号楼
    延安西路 221 号（近乌鲁木齐北路）
    静安区

39. 爱文义公寓 (1931-1932)
    今联华公寓
    北京西路 1341-1383 号（近铜仁路）
    静安区

49. 吴同文住宅 (1935-1938)
    今上海规划师之家和上海城市规划
    博物馆（改建中）
    铜仁路 333 号（近北京西路）
    静安区

Zone C

1. The 22 Residences on Route Ratard (1919-1920)
   No. 1-10 Lane 852 Julu Rd,
   No. 868-892 Julu Rd (near Changshu Rd), Jing'an District

5. Jean Beudin's Residence (1919-1920)
   now "Ambrosia" BBQ
   No.150 Fenyang Rd (near Taojiang Rd)
   Xuhui District

15. Country Hospital (1923-1926)
    now No. 1 Building of
    Huadong Hospital,
    No. 221 West Yan'an Rd (near Middle Wulumuqi Rd), Jing'an District

39. Avenue Apartments (1931-1932)
    now Lianhua Apartments
    No. 1341-1383 West Beijing Rd
    (near Tongren Rd), Jing'an District

49. D.V.Woo's Residence (1935-1938)
    now Shanghai Urban Planner's Saloon and Urban Planning Museum (under construction),
    No. 333 Tongren Rd (near West Beijing Rd), Jing'an District

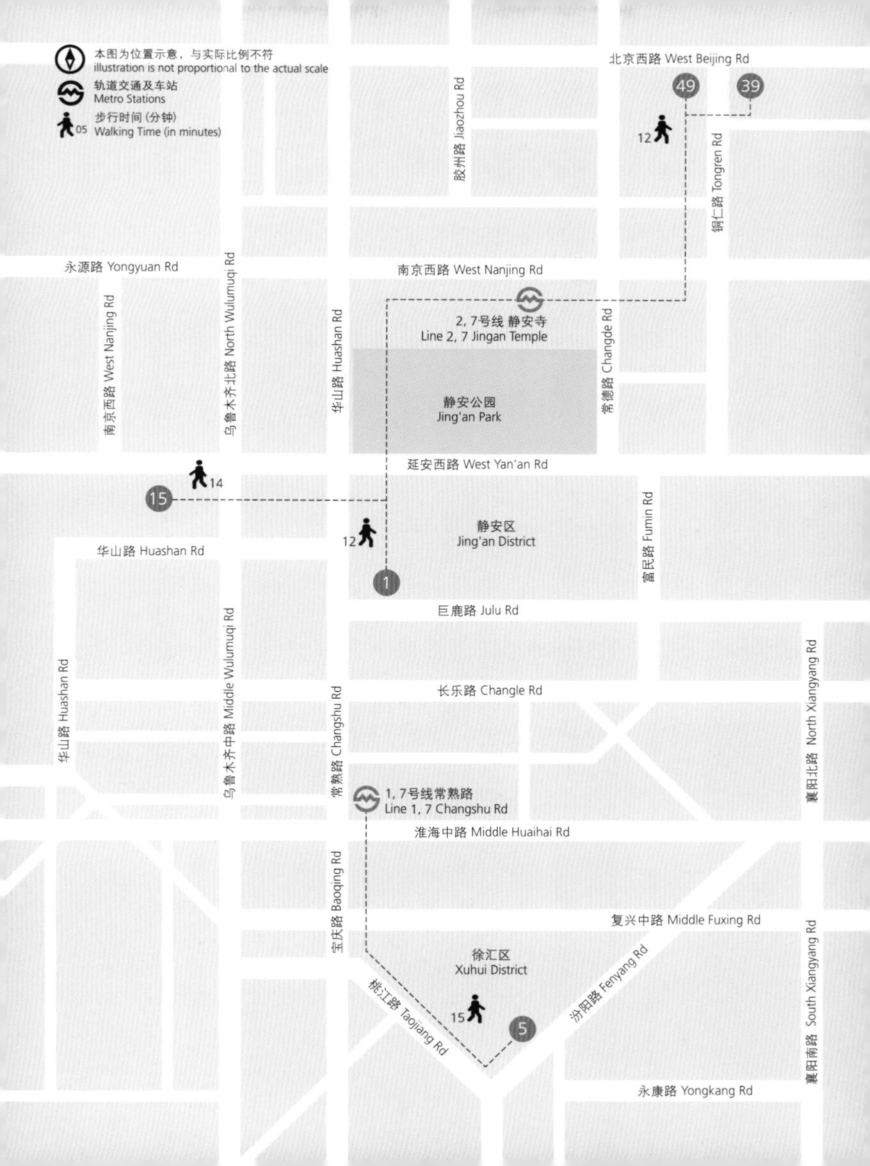

本图为位置示意，与实际比例不符
illustration is not proportional to the actual scale
轨道交通及车站
Metro Stations
步行时间（分钟）
05
Walking Time (in minutes)
北京西路 West Beijing Rd
49
39
12
胶州路 Jiaozhou Rd
铜仁路 Tongren Rd
永源路 Yongyuan Rd
南京西路 West Nanjing Rd
南京西路 West Nanjing Rd
乌鲁木齐北路 North Wulumuqi Rd
华山路 Huashan Rd
2, 7号线 静安寺
Line 2, 7 Jingan Temple
常德路 Changde Rd
静安公园
Jing'an Park
延安西路 West Yan'an Rd
14
15
12
静安区
Jing'an District
富民路 Fumin Rd
华山路 Huashan Rd
1
巨鹿路 Julu Rd
华山路 Huashan Rd
乌鲁木齐中路 Middle Wulumuqi Rd
常熟路 Changshu Rd
长乐路 Changle Rd
襄阳北路 North Xiangyang Rd
1, 7号线常熟路
Line 1, 7 Changshu Rd
淮海中路 Middle Huaihai Rd
宝庆路 Baoqing Rd
复兴中路 Middle Fuxing Rd
徐汇区
Xuhui District
汾阳路 Fenyang Rd
桃江路 Taojiang Rd
15
5
襄阳南路 South Xiangyang Rd
永康路 Yongkang Rd

巨籁达路 22 栋住宅
The 22 Residences on Route Ratard

邬达克在上海建成的第一个项目
Hudec's First Completed Project in Shanghai

邬达克的诸多业主中，有两个民国时期影响很大的储蓄会，一个是投资国际饭店的中商四行储蓄会，另一个是法商万国储蓄会。

万国储蓄会于 1912 年 9 月由法国商人在上海成立，是旧中国最早也是最大的储蓄机构，其名号表明该会对投资人没有国籍限制。主要业务是以有奖储蓄吸收居民小额存款，积聚资金主要用于投资上海房地产和外资企业股票与债券，利润丰厚，曾在多座城市设立分会。1934 年，有奖储蓄被取缔后，影响才逐渐减弱直至消除。

邬达克至少为万国储蓄会设计过 3 个重要项目：原爱多亚路（今延安东路 7-9 号）万国储蓄会大楼和方西马大楼（1997 年建延安路高架时拆除）、巨籁达路（今巨鹿路）22 栋住宅和霞飞路（今淮海中路）诺曼底公寓。前两个项目是邬达克初到上海时，在克利洋行设计的早期作品。

今天，相对于备受关注的诺曼底公寓，巨籁达路住宅成为隐藏在花园深处鲜为人知的邬达克作品。然而 90 多年前，这批较早在上海法租界实施的美式风格住宅却曾广受欢迎。刚建成时，就有多家美资公司租赁房屋作为员工宿舍。

22 栋独立住宅总体布局构成 L 形，其中 14 栋朝向巨籁达路一字排开，其余 8 栋则分列在支弄两旁。整个别墅区事实上只有两种标准户型：底层设门廊、进厅、起居室、餐厅、配餐室和厨房，二层有三个卧室，两个卫生间和一个凉廊，阁楼上还有两个房间以及佣人房。房间铺设橡木地板，室内还配备美国式的集中供暖系统，锅炉房放在地下室。

为了避免同样的平面造成千篇一律的外形，设计师煞费苦心地设计了 4 种立面风格，变化主要在坡屋顶、凉廊、入口和材质细部，住宅群整体构成统一而丰富的外观。每栋别墅前还有大片开敞的草坪。

邬达克在现场监理完成了该项目的施工。因为孤身一人，初到上海，思乡心切，他几乎每天写一封家书，向亲人介绍自己的工作和生活。作为邬达克在上海建成的第一个项目，巨籁达路 22 栋住宅就是其中介绍最详细的内容之一。

这 22 栋住宅现为巨鹿路花园住宅，1999 年被分别列入上海市优秀历史建筑保护名录。

参观指南

建筑内部不对外开放，沿巨鹿路可以欣赏部分外观，支弄也可进入。

CATV
4.

Hudec had served for two influential banking institutions in Shanghai. One was the Joint Savings Society which was founded by four Chinese banks, and the owners of the Park Hotel. The other was International Savings Society with French investment.

The International Savings Society was founded in Shanghai by French merchants in September of 1912. The name implied that there was no limitation over the nationality of an investor.

The society attracted small-amount of savings from local residents and received handsome profits from investing the money on the city's real estate or stocks and bonds of foreign enterprises. During its peak, the society had branches in many cities but its influence was limited after prized savings were prohibited in 1934.

Hudec had designed at least three important projects for the International Savings Society, including its headquarters on No. 7 Avenue Edward VII (which was demolished when the Yan'an Road elevated highway was being built), 22 Residences on Route Ratard (Julu Road) and the Normandie Apartments (now Wukang Building) on Avenue Joffre (today's Middle Huaihai Road). The previous two projects were his very early works in Shanghai when he still worked for the R.A. Curry firm.

Compared with the eye-striking Normandie Apartments, the 22 residences on Julu Road are hidden behind deep lanes and gardens and rarely known.

Some 90 years ago or beyond, trendy American-style houses in the French Concession were immensely popular, which were rent by American companies as residences for their employees.

The 22 residences were arranged in the shape of a big "L," 14 of which were row houses facing the Julu Road. The remaining eight houses were flanked on both sides of a side lane.

The villas were designed according to two standard layouts, with a porch, an entrance hall, a sitting room, a dining hall, a kitchen on the first floor and three bedrooms, two bathrooms and a loggia on the second. The loft also featured two rooms and the servants' rooms.

The rooms were equipped with oak wood floorings and American heating system. The boiler room was located in the basement.

Hudec had designed façades in four styles to avoid uniform appearances, which varied from the sloping room, the loggia, the entrance and the details of textures. Therefore he created an integrated yet varied effects for the group of 22 buildings. Every villa was fronted with spacious lawns.

Hudec supervised the construction on the site. The man who came to Shanghai all by himself was home sick. He wrote a letter to his family almost every day, talking about his work and life in a foreign city. As his first completed project in Shanghai, the 22 villas were among the most detailed mentioned work in his letters.

The 22 villas on Julu Road were listed as Shanghai Excellent Historical Buildings in 1999.

Tips

The villas are not open to the public but the façades can be admired from Julu Road.

盘滕住宅 Jean Beudin's Residence

从董事府邸到白公馆再到仙炙轩
From Beudin's Residence to Pai's Mansion to Ambrosia

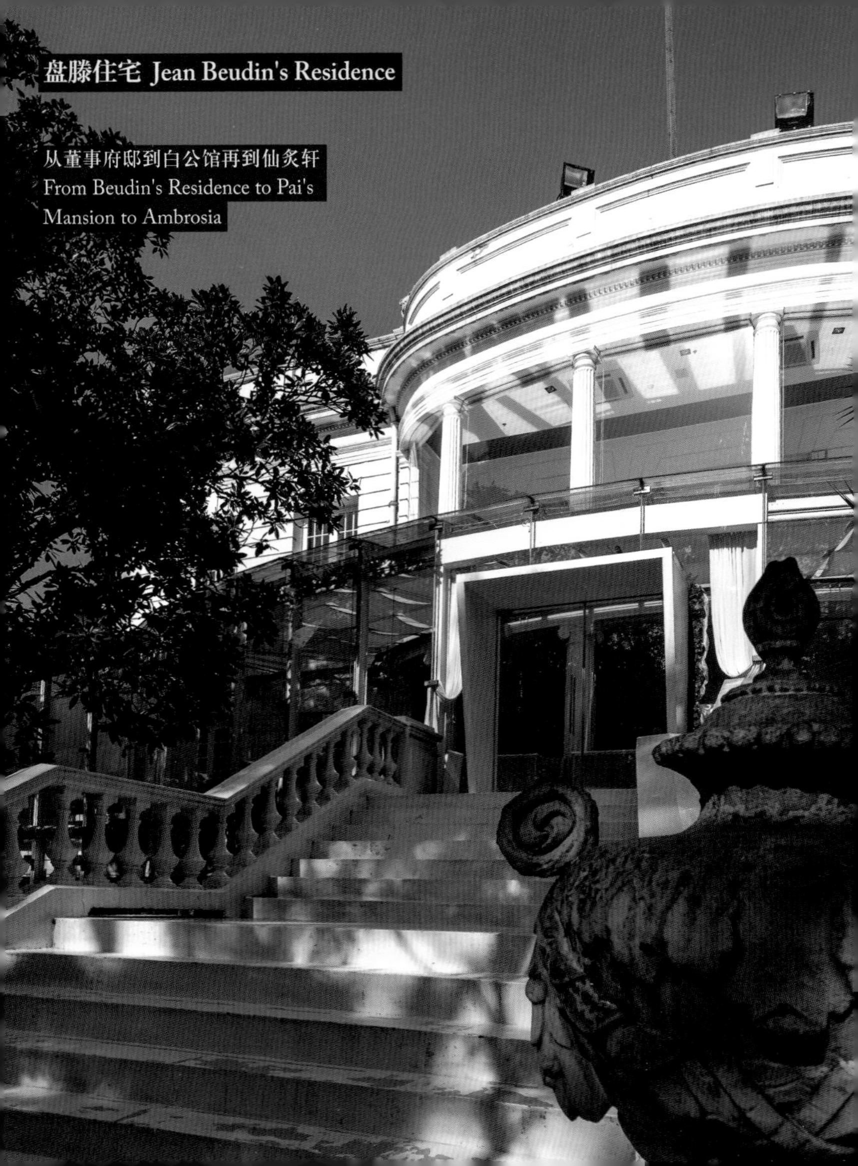

作为近代上海的建筑明星，邬达克的作品常伴有专业及大众媒体的广泛报道，因此其公共建筑大部分为人所熟知，私宅的设计归属却存在较大争议。一方面出于安全考虑，私宅鲜有曝光；另一方面，其设计投入和艺术成就可能也不及大型作品。因此，除像“绿房子”这样少数经典外，考证邬达克操刀的其他私宅颇具难度：那些 1920 年前后为外商设计的豪华公馆，年代久远，信息不明确；而其自行开业后为地产公司设计的商品住宅，则数量众多，位置分散，特征又不明显。

原法租界毕勋路（今汾阳路）的两幢古典主义白色府邸，因为设计精美、地位重要而成为中外学者争议的焦点。因为邬达克后人拥有这些建筑的蓝图、照片和相关信件，但图纸没有签名，因此尚需其他佐证，如原法租界公董局总董马迪耶府邸（汾阳路 79 号，曾为陈毅府邸，今上海工艺美术博物馆）。而汾阳路 150 号则已被多方证实确系邬达克在克利洋行时期为万国储蓄会董事盘滕（或译为部亭）设计的住宅。

这是一座法国文艺复兴风格的二层混合结构建筑，另有半地下室一层构成宽大的基座，立面为横三段竖三段处理。顺着大楼梯拾级而上，是两层高的弧形大门廊，首层采用爱奥尼柱式，二层为塔斯干柱式支撑的阳台，女儿墙为饰有宝瓶的石栏杆。墙面分缝明显，窗上有楣饰，建筑通体为白色。

虽然建筑的正立面和花园的水池都对称设置，但平面却较为灵活。圆形主楼梯突出北立面，与东侧半圆形大厅及依附一旁的弧形室外楼梯形成呼应。室内一层为公共区域，包括门厅、客厅和餐厅。二、三层南面为主人卧室、起居室，北面为佣人房，地下室包含厨房、设备、佣人房和储藏间。

1945 年，万国储蓄会破产清算后，名下房产为政府收购。这栋豪宅成为国民革命军参谋长、桂系首领白崇禧的住宅，故人称“白公馆”。

新中国成立后，该楼先归上海国画院，后为越剧院的办公室和排练场，名誉院长袁雪芬的办公室就是白崇禧之子白先勇 1948 年离沪前居住的房间。

2004 年，当年的“白公馆”半圆形露台已用落地玻璃封闭，变身为一家高档的日式烤肉餐厅——“仙炙轩”，与上海第一家“宝莱纳”啤酒餐厅共享着入口和花园。

参观指南

花园内可欣赏建筑立面，室内需用餐才可参观。

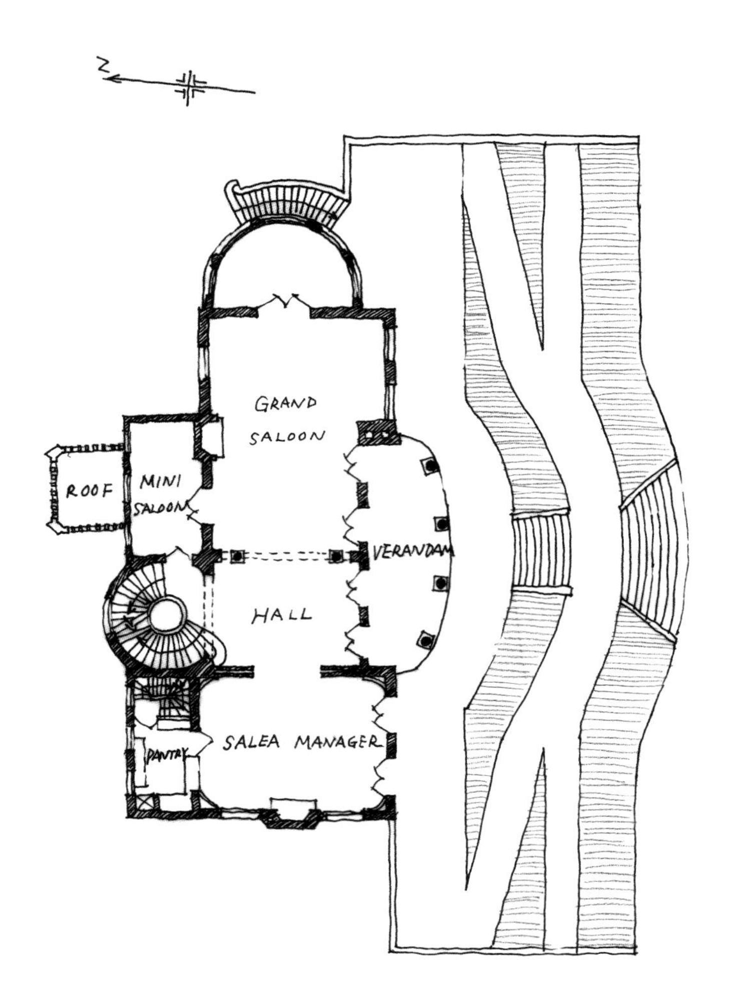

首层平面 Ground Floor Plan

In the 1980s, Shanghai Yueju Opera House invited Taiwanese writer Pai Hsien-yung to dine at a Chinese Restaurant on the quiet Fenyang Road (formerly Route Pichon), which was lined with musical stores and lush Chinese parasol trees. It was on this night Pai discovered the familiar surroundings that used to be his home.

According to recent archival findings, the beautiful white villa on No. 150 Fenyang Road is another Hudec masterpiece. He designed the residence for Jean Beudin, who was the director of the International Savings Society when he worked at R. A. Curry's firm.

The two-story villa in composite structure is designed in French Renaissance style and sits on a huge base. The façade is divided in classic three sections — both vertically and horizontally. A giant staircase leads up to a curved veranda which is as high as two stories.

The ground floor is adorned with Ionic orders while the balcony of the second floor is supported by Tuscan orders. The parapet is graced by vase-shaped stone balusters. The villa is painted in snow-white color.

Hudec had created a vivid layout for the building. The round-shaped main staircase on the northern facade echoes with the semi-circled hall on the eastern side and a neighboring curved outdoor staircase.

The ground floor is mainly public area which includes an entrance hall, a sitting room and a dining room. The master bedrooms and sitting room are arranged on the southern side of the second and third floors while servants' rooms are on the northern side. The basement is equipped with a kitchen, a store room, etc.

The Kuomintang government took over the building in 1945 after the International Savings Society went bankrupt. Kuomintang general Pai Chung-hsi — father of the renowned writer Pai Hsien-yung — moved into the house. Since then, the white villa has been known to the world as "Pai's Mansion."

After the Pais moved to Taiwan in 1949, the villa has been home to the China Art Academy and later the Shanghai Yueju Opera House. The Opera House had rented out a portion of the house to a Chinese restaurant.

In 2004, a Japanese BBQ restaurant "Ambrosia" enclosed the curved veranda with glasses. It shares the garden with the city's first Paulaner Brauhaus restaurant.

It is still unclear if another villa on Fenyang Road — the Madier Residence on No. 79 — was also built by Hudec. Although Hudec's descendants showed the blueprints, pictures and relevant letters of the building, the blueprints did not carry his signatures. This house was the home of Shanghai's first mayor Chen Yi after 1949. The Shanghai Arts and Crafts Research Institute took over the place in 1962 and in 2001, started a museum to showcase traditional Chinese handcrafts.

Tips

The façade can be admired from the garden. The interior is open to restaurant's clients.

宏恩医院 Country Hospital

独立开业后完成的第一件重要作品
Hudec's First Important Project in His Independent Career

1923年末，宏恩医院的神秘捐赠者，一对美国富商夫妇，因为在上海致富又无子嗣继承，希望回馈上海，造福国际社区。他们委托邬达克设计一座医院，造价没有限制，但若泄露其身份，合约无条件终止。

宏恩医院定位更接近一家疗养院。选址在公共租界与法租界交界线的延伸段，到外滩开车仅需10分钟，交通便利，往西是大片农田，环境幽静。

基于上海的气候，五层病房楼朝南一字排开，面向花园，医务楼临街布置在北侧，大楼的平面整体呈“工”字形。标准病房配备进口三件套的独立卫生间，整座大楼装载瑞士苏尔寿的冷气系统，这在上海系首例，全世界亦屈指可数。医疗设施也是全上海最大最好的。

建筑南立面采用意大利文艺复兴风格，以获取国际社区普遍的文化认同。三组顶部饰有山墙的巨大凉廊，主要用来遮挡夏季毒辣的阳光。根据统计，当时上海夏季病房的入住率约为60%，凉廊在立面上的比例就此确定。

参观者一踏进医院门厅，就感觉像进入了一个华美的旅馆大堂。这里没有压抑冰冷的医院气氛，反而让人联想到托斯卡纳地区的乡村风光，这都源于捐赠者早年久居意大利留下的情结。门厅和接待大厅所有的大理石、雕饰和家具也都从意大利进口。

工程伊始，捐赠者就开始规划、种植和养护花园，草坪尽头是广玉兰，四周是本地生长的各种灌木和花卉。3年后，迎接第一位病人的不是满地瓦砾，而是宜人美景。

宏恩医院建成后被誉为“远东最好的医院”，国内外媒体争相报道，时任工部局总董的费信惇应邀担任医院管理委员会主席。

这是邬达克独立开业后完成的第一件重要作品，从此，这位“上海有前途的知名建筑师”开始跻身“远东最前沿建筑师的行列”。因为设计师信守诺言，1926年初夏，捐赠者得以亲历自己赠予上海的这份厚礼的落成仪式。

1950年10月，宏恩医院由人民政府接管，次年8月更名为华东医院。原大楼现为华东医院1号楼，即老干部病房楼。这里是上海高干高知看病的地方，著名作家巴金先生曾在此楼度过生命中最后6年的时光（1998–2005）。

参观指南

建筑不对外开放。

大厅设计草图 Sketch of Hall

Buildings are often built from dreams. For an heirless American businessman living in Shanghai in the 1920s, the dream was to build a luxurious hospital for expats with central air-conditioning and a bathroom in each ward.

The man who had made his fortune in Shanghai, donated funds to build the Country Hospital in 1926, which is the No. 1 building of today's Huadong Hospital on Yan'an Road which looks after senior citizens.

The building sits on a huge garden dotted with century-old pines and magnolia trees while elderly people move about in wheelchairs.

The businessman asked Hudec to design the high-end hospital for Shanghai expatriates. He insisted on the highest standards and the best facilities, and there was no budget limit — a perfect project for an architect. The benefactor also asked Hudec to keep his identity a secret, or the contract would be canceled.

Covering 2,300 square meters, the five-story steel-concrete building is perched near Yan'an Road inside the former French Concession, only 10 minutes' drive from the Bund.

As one of Hudec's early works, the building is in classical style, different from his later modern works. The hospital's southern façade features a tripartite vertical composition with balanced east and west wings.

The ground floor consists of double-columned arched windows and a continuous arcade, a design to prevent strong summer sunlight. The second to fourth floors are treated with a unified approach. The fifth floor and the parapet are designed as three imposing pediments and make a powerful impression on visitors. Fortunately there was air-conditioning, because the temperature inside the operating room in Shanghai could reach 42 degrees Celsius in summer.

The ground floor looks more like a stylish hotel lobby — quiet and elegant — rather than anything like a hospital entrance. Hudec made extensive use of imported black, white and gray marble in classical columns, large staircases and flooring.

The patterned wooden ceiling makes an interesting textural contrast with the smooth marble as the sunlight pours into the lovely lobby from a row of large French windows. The bright and open lobby totally changed the usual cold, dim decor of a hospital.

On completion, it was considered the best hospital in the Far East with the *Shanghai Sunday Times* newspaper commenting: "The rural, restful early Italian Renaissance effect reminds visitors of the courtyard and rooms of a villa in Tuscany."

Hudec chose the Italian Renaissance style because the mysterious benefactor had spent much of his life in Italy.

Since Hudec had obeyed the agreement, the philanthropist is said to have observed with satisfaction that he enjoyed "standing in the front row of the opening ceremony with no one knowing that I built the hospital."

Tips

The building is not open to the public.

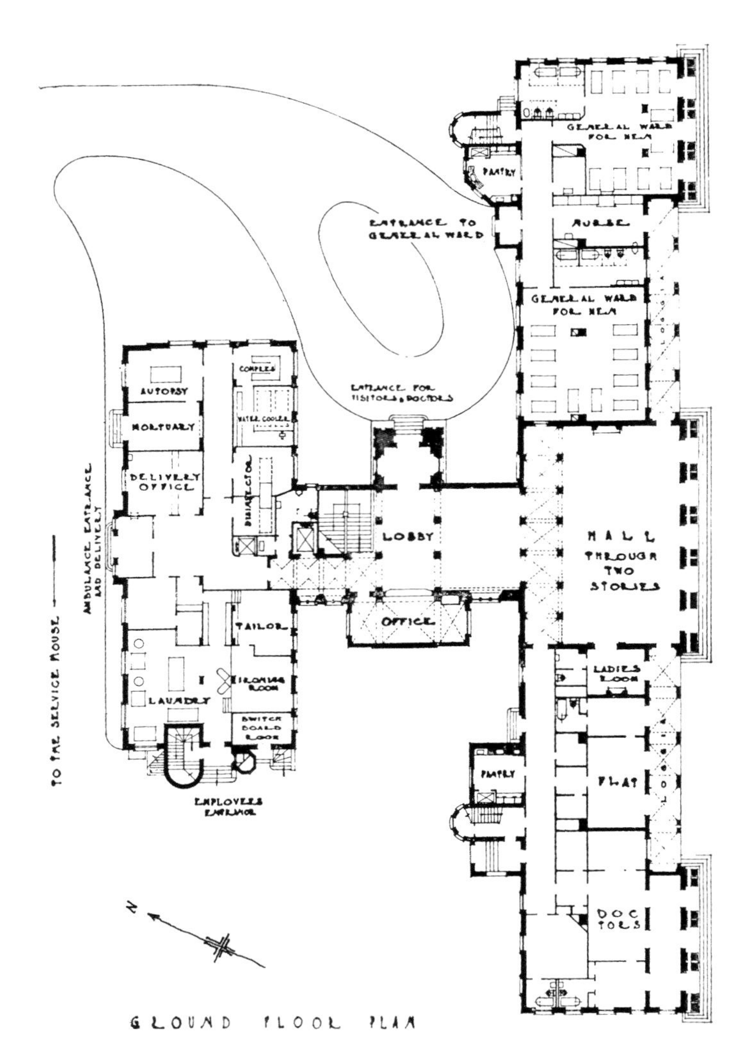

首层平面 Ground Floor Plan

爱文义公寓 Avenue Apartments

爱文义路上的多层公寓
Multistory Apartments on the Avenue Road

在今天铜仁路（原哈同路）北端与北京西路（原爱文义路）相交的丁字路口两侧，矗立着两个邬达克设计的项目，西侧是著名的吴同文住宅，东侧是一组多层板式公寓——爱文义公寓，1943年后更名为联华公寓。

爱文义公寓地处当时公共租界的边缘区，由联合房地产公司投资兴建，1931年3月设计，1932年5月建成。当时，因为租界人口和房产价格的一路飙升，各种住宅大楼在建成前通常已被大量预定。然而爱文义公寓的情况更为稀罕：项目尚未完成立项，96%的公寓已被抢租一空，并且来预订的不少是熟悉该工程设计者，包括参与建设和审批的技术专家和公务员。项目的吸引力主要来源于精致的功能布局、周围便利的交通、邻近的公共活动场地，并且只要支付少量租金就可以使用车库和电冰箱。

需要指出的是，今天的联华公寓包括三栋五至六层的板式住宅楼，这与邬达克最初的规划不尽相同。1931年规划的爱文义公寓只有两栋四层住宅楼，包含56套公寓。北面沿街是13个两层楼的中国商店（一层店铺，二层为店主居室）。住宅和商店之间是19个车库和一个锅炉房。今天沿北京西路令人瞩目的六层弧形公寓和原来两幢楼的加层是何时何人所为，还有待进一步查证。

在最初的设计中，四层公寓楼每层各有四个单元，户型从一室户到五室户不等。卧室有全套卫生设备，统一供应煤气及热水。缺乏自然通风和采光的卫生间利用风井拔风。室内，新加坡红木铺地，储物间布置合理，壁橱嵌入墙内。公寓设有南北两套通行路线，业主由南入口进入，楼梯宽敞堂皇，仆佣则从北入口进入，楼梯狭窄紧凑。

建筑整体外观为现代风格，立面简洁，窗间为红色清水砖墙，层间为水刷石，米色与砖红色交替形成强烈的水平线条。北面楼梯作竖向构图，南侧公寓沿哈同路形成流畅的弧面。阳台栏板中间和楼梯栏杆的铸铁花饰为装饰艺术风格。

1999年，爱文义公寓被列入上海市优秀历史建筑保护名录。

参观指南

沿北京西路、铜仁路、南阳路可以参观建筑外观，室内不对外开放。

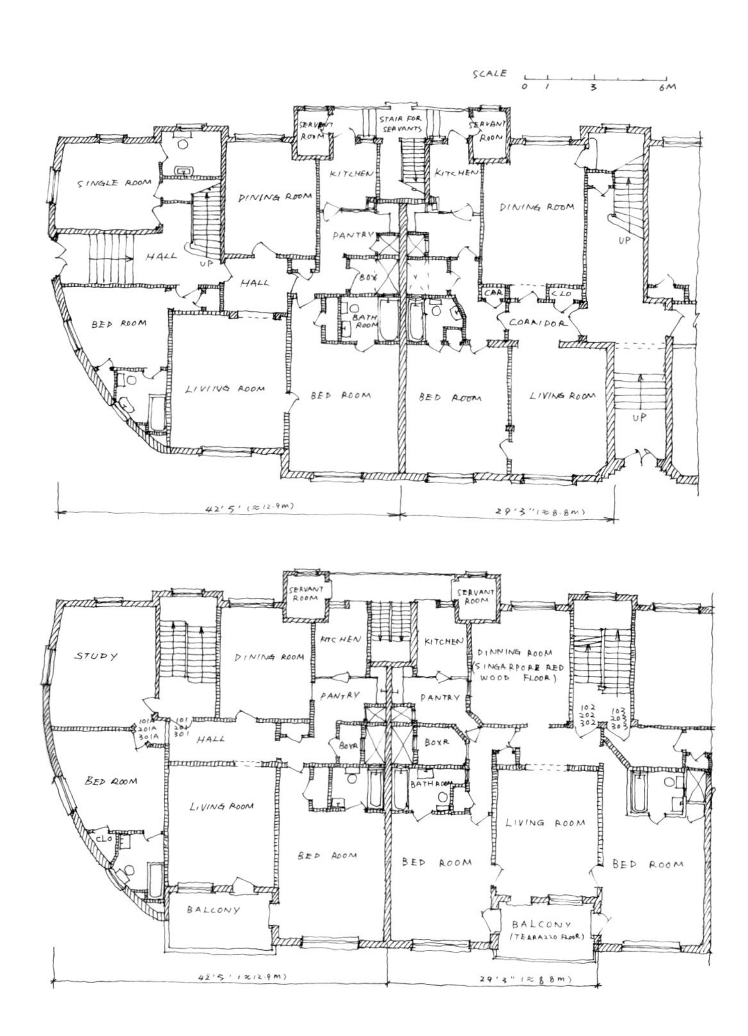

一层、二层平面 Ground Floor Plan and 1st Floor Plan

On the crossroad of Tongren Road (former Hardoon Road) and West Beijing Road (former Avenue Road) stand two Hudec projects. Famous tycoon Wu Tongwen's residence, nicknamed "Green Villa", sits on the west side while on the east stands the multistory Avenue Apartments, known today as Lianhua Apartments.

Situated on the edge of the former International Settlement, Union Real Estate Co. invested in Avenue Apartments. It was designed in March 1931 and completed the following year.

The project was built around the time when both the population and real estate prices in the settlement area were skyrocketing. Some apartments were booked even before the completion of the project.

The Avenue Apartments were even more popular. Around 96 percent of the suites were leased out before the project even kicked off. Many tenants were participating technicians or civil servants, who were familiar with the project. They were attracted by the functional layout, the convenient transportation and adequate facilities for public activity in the neighborhood. The permission to use the garage and refrigerators with only a small amount of money was an added bonus.

Today the Apartments include three residential buildings with five to six floors, which are different from the original design. The 1931 plan showed only two four-story apartments containing 56 suites. A number of 13 two-story Chinese shops lined on the northern façade fronting the street. The boiler room and 19 garages were installed between the residence and the shops. It is still unclear who built the eye-catching six-story curved-shaped apartment building on the West Beijing Road and added two more stories later on.

In the initial design, each floor of the four-story apartments was divided into four sections. Every section contained suites ranging from one to five bedrooms, which was equipped with bathroom, gas cooker and hot water supply. A ventilating shaft had been used to provide fresh air for the bathrooms lacking natural ventilation and lighting.

The rooms were paved with Singapore mahogany flooring and filled with functional facility for storage including in-wall closets. The architect had built two sets of passenger routes. The owners could walk in from the southern entrance with grand staircases while their servants entered through the northern gate with a narrow staircase.

The buildings are essentially modern in style with a simple-cut façade. Hudec had used red-brick walls between the windows and rustic finish between the stories. A strong visual effect is created by the contrasting creamy and red colors.

The northern staircase is in a vertical design while the southern façade shapes like an arc on Hardoon Road. The balustrades over the balconies and staircases are graced by Art-Deco cast iron patterns.

The Avenue Apartments were listed the Shanghai Excellent Historical Buildings for Preservation in 1999.

Tips

The buildings can be appreciated on the West Beijing Road, Tongren Road and Nanyang Road. The interior is not open to the public.

吴同文住宅 D.V.Woo's Residence

远东最大最豪华的住宅之一
One of the Largest Luxurious Residences in the Far East

吴同文住宅是邬达克设计成熟期的代表作，上海经典的现代风格花园洋房。它曾是“远东最大最豪华的住宅之一”，建筑面积近 1700 平方米，因为外立面和围墙均采用绿色釉面砖，俗称“绿房子”。

吴同文是旧上海著名的颜料商，也是显赫的苏州籍巨商贝润生的女婿，因军绿色颜料致富，故视绿色为幸运色。该地块是贝家小姐的陪嫁礼，毗邻的哈同路和爱文义路恰好蕴含了主人的名字，而门牌 333 号据说系主人重金购买，代表占地 3 亩 3 分 3 厘。

1935 年，国际银价危机拖累上海经济，房地产项目大多停滞。邬达克洋行从巅峰坠入谷底，生存堪忧，吴宅的委托几成救命稻草。或许也因为巨大的压力和相对充裕的时间和精力，设计几易其稿，不断精进。

该宅充分体现了有机建筑的设计原则：布局精密紧凑，与基地完美契合。主体紧贴北侧道路，与顺应转弯半径的弧形围墙连成整体。首层中间架空作汽车道，压缩交通面积，同时把功能分成两部分：西式的社交空间，包括酒吧、弹子房、餐厅等和上层主人用房向南面花园开敞，北面的中式客堂、祖屋和佣人房较封闭。半圆形楼梯外通高的玻璃正对道路转角。

南立面设计精湛：圆柱形的阳光房通高四层，与层层退进，曲线流畅的大露台形成纵横对比，顺餐厅外墙盘旋而上的弧形大楼梯将露台与花园连为一体。楼梯与阳台的铸铁花饰为简洁的现代风格，阳光房外均采用进口的圆弧玻璃，甚至连玻璃移门也是弧形的，且至今活动自如。

室内装修豪华：日光室上覆玻璃顶棚，小舞厅安装弹簧地板，煤卫冷暖设备齐全，并首次在上海私宅中安装了电梯，还是独特的荷叶形平面。车道两侧和楼梯墙面采用意大利洞石，墙面铸铁花饰为装饰艺术风格（今已不存）。一层的佛堂和祖屋则完全是中式风格。

新中国成立后，该楼一、二层被捐作上海工商联活动俱乐部。“文革”初期，吴同文和姨太太在自己钟爱的绿房子内自尽。

绿房子的传奇后成为作家程乃珊的小说《蓝屋》的原型。

1978 年后，绿房子划归上海市规划设计研究院做办公楼，后改为晒图室。2003 年由台商租赁，改为餐厅酒吧等。近期，规划院已收回房产。保护更新后，这里将成为上海规划师之家和上海城市规划博物馆，近百年来改变上海城市面貌的那些珍贵蓝图将可能与公众见面。

参观指南

沿铜仁路和北京西路可以欣赏建筑外立面。改造完成后，建筑首层将对公众开放。

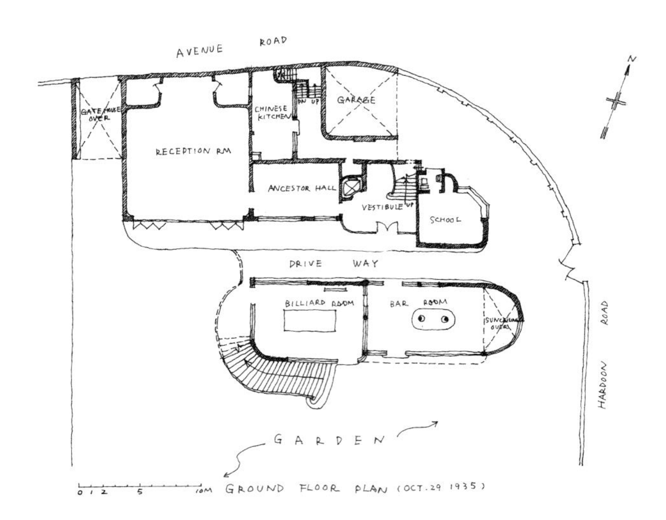

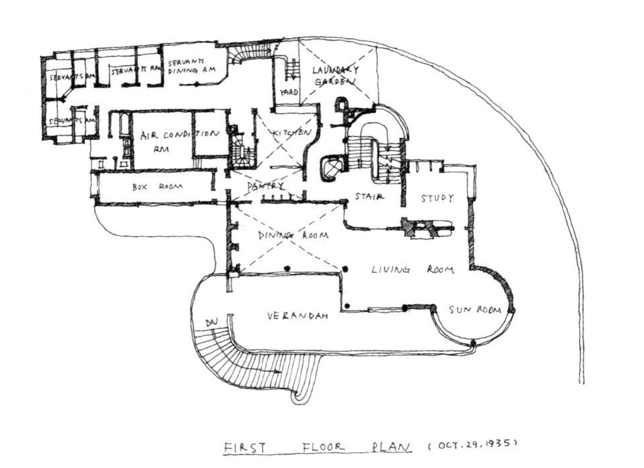

首层、二层平面 Ground Floor Plan and 1st Floor Plan

D. V. Woo's Residence is a signature Hudec work in his prime. This modern-style garden villa was formerly the largest and most luxurious residence in the Far East. Covering an area of 1,700 square meters, the villa was nicknamed "green house" because of green-colored glazed tiles on the façade and surrounding walls.

D. V. Woo was a famous merchant who sold pigments in old Shanghai and also happened to be the son-in-law of Suzhou tycoon Bei Runsheng. Mr. Woo made his fortune by selling green pigments for military use and regarded green as his lucky color. The site was a dowry from his wife.

Shanghai's economy was dragged down by international silver crisis in 1935, which halted many real estate properties. Hudec's firm, riding at the top, almost had a freefall when Mr. Woo commissioned for the work. It was a true savior. Due to huge pressure but with more free time on hand, Hudec changed the plans several times to make it a perfect house.

The residence showcases the essence of organic architecture. It has an accurate and concise layout, which merges perfectly with the base. The main body approaches the road on the north, which connects with the curved surrounding walls on the corner as a unity.

The center of the ground floor was designed as passage for cars through which Hudec has made the land for transportation as compact as possible.

He also divided the building into two parts according to functions. The western-style social space including a bar, a billiard room, a dining hall and main bedrooms face an open garden. The Chinese-style sitting hall, the ancestral room and servants' room on the northern side are in a more closed style.

The southern façade is stylish. The cylinder-shaped sun parlor is four floors high, which contrasts with the big balcony full of smooth curves that is shrinking layer by layer. The cast iron patterns on the staircase and balcony are in modern style. The sun parlor is made with imported curved glass. Even the glass door is in a curved shape.

The interior is embellished in luxurious style. The sun parlor is topped with a glass ceiling. The dancing hall is equipped with sprung floors. The villa is installed with gas supply, bathrooms and cooling and heating systems. An elevator in a unique lotus leaf shape was the first to be installed in any private residences in Shanghai.

The walls on both sides of the car passage and beside the staircase are built with Italian cave stones. The Buddhist room and ancestral room are in a pure Chinese style.

The ground and first floors were used as the club for Shanghai Association for Industry and Commerce after 1949. D. V. Woo and his mistress committed suicide in their favorite "green house" at the start of the Cultural Revolution (1966-76).

The "green house" was used by the Shanghai Urban Planning and Design Institution as its office after 1978. A Taiwanese merchant renovated the building as a restaurant & bar in 2003. However, the Urban Planning institution has taken it back and is renovating it as a club and museum for urban planners. Precious blueprints which had changed the urban scene of Shanghai might be exhibited in the "green house".

Tips

The façade can be appreciated from Tongren and West Beijing Road. The ground floor will be open to the public after the renovation.

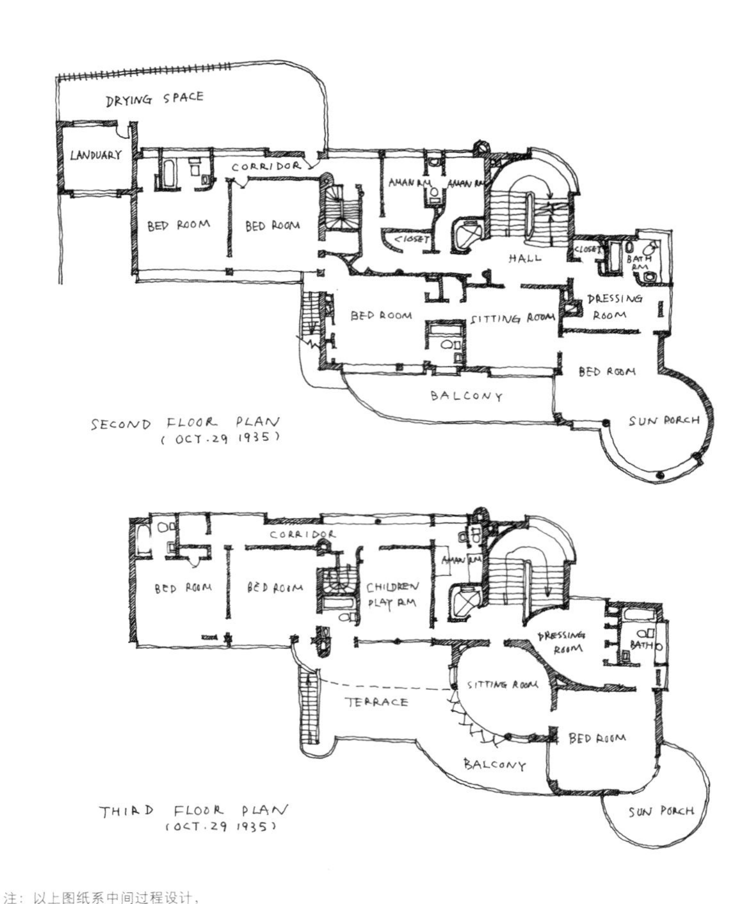

注：以上图纸系中间过程设计，
后有数次修改，调整了车库、
围墙门卫和平台楼梯。
Note: The draft for the garage,
walls and staircase was revised later

三层、四层平面 2nd Floor Plan and 3rd Floor Plan

D 区

13. 诺曼底公寓 (1923-1926)
今武康大楼
淮海中路 1836-1858 号（近武康路）
徐汇区

27. 邬达克自宅 (1930)
今邬达克纪念室（底层）
番禺路 129 号（近平武路）
长宁区

30. 孙科住宅 (1929-1931)
今“万科之家”（改造中）
延安西路 1262 号（近番禺路）
长宁区

33. 交通大学工程馆 (1931)
华山路 1954 号（近番禺路）
徐汇区

34. 哥伦比亚住宅圈 (1928-1932)
今外国弄堂私人住宅
新华路 119、155、185、211、236、248、276、294、329 弄部分住宅
（近番禺路）
长宁区

48. 达华公寓 (1935-1937)
今上海锦江都城达华酒店
延安西路 918 号（近江苏路）
长宁区

Zone D

13. Normandie Apartments (1923-1926)
now Wukang Building,
No. 1836-1858 Middle Huaihai Rd
(near Wukang Rd), Xuhui District

27. Hudec's Residence (1930)
now Hudec Memorial Hall (ground floor),
No. 129 Panyu Rd (near Pingwu Rd),
Changning District

30. Sun Ke's Residence (1929-1931)
Now Shanghai VankeHouse
(Under Renovation)
No. 1262 West Yan'an Rd
(near Panyu Rd), Changning District

33. Engineering and Laboratory Building of Chiao Tung University (1931)
No. 1954 Huashan Rd (near Panyu Rd),
Xuhui District

34. Columbia Circle (1928-1932)
now Foreigner's Lane Garden Villa,
Some Villas of Lane 119, 155, 185, 211, 236, 248, 276, 294, 329 Xinhua Rd
(near Panyu Rd),
Changning District

48. Hubertus Court (1935-1937)
now Shanghai Jinjiang Metropolo Dahua Hotel,
No. 918 West Yan'an Rd
(near Jiangsu Rd),
Changning District

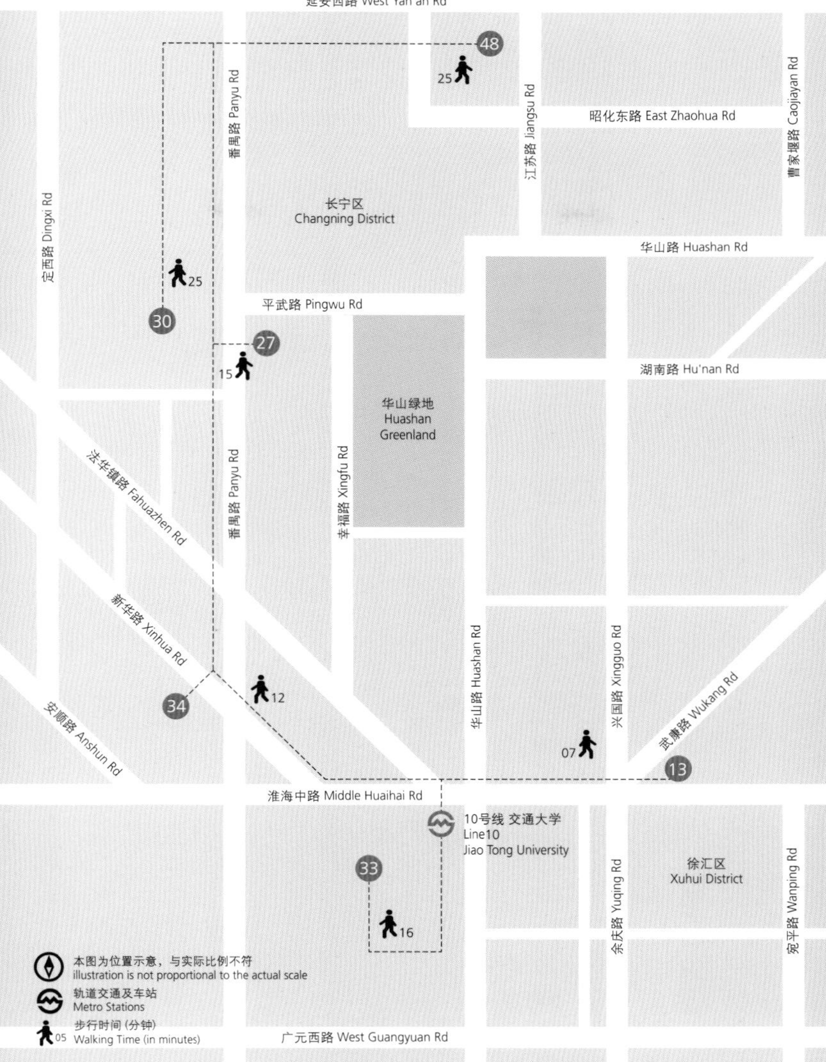
延安西路 West Yan'an Rd
48
25
番禺路 Panyu Rd
江苏路 Jiangsu Rd
昭化东路 East Zhaohua Rd
曹家堰路 Caojiayan Rd
长宁区
Changning District
定西路 Dingxi Rd
华山路 Huashan Rd
25
30
平武路 Pingwu Rd
27
15
湖南路 Hu'nan Rd
华山绿地
Huashan
Greenland
法华镇路 Fahuazhen Rd
番禺路 Panyu Rd
幸福路 Xingfu Rd
新华路 Xinhua Rd
华山路 Huashan Rd
兴国路 Xingguo Rd
12
34
武康路 Wukang Rd
安顺路 Anshun Rd
07
13
淮海中路 Middle Huaihai Rd
10号线 交通大学
Line10
Jiao Tong University
33
余庆路 Yuqing Rd
徐汇区
Xuhui District
宛平路 Wanping Rd
16
本图为位置示意，与实际比例不符
illustration is not proportional to the actual scale
轨道交通及车站
Metro Stations
步行时间 (分钟)
05 Walking Time (in minutes)
广元西路 West Guangyuan Rd

诺曼底公寓 Normandie Apartments

沪上最早的外廊式公寓
The Earliest Veranda-Style
Apartments in Shanghai

万国储蓄会霞飞路公寓亦称诺曼底公寓或东美特公寓。这是邬达克在克利洋行时期的重要作品，也是为万国储蓄会设计的作品中最有影响的一个。

对于该公寓的命名，如今争议颇多。从大楼建成时间来看，纪念“二战”诺曼底登陆是不可能的，纪念“一战”中被击沉的诺曼底号战舰之说也有待查实，更合理的解释是以法国地名命名，恰好纪念了这个朝向英吉利海峡的地区。理由是万国储蓄会下辖中国建业地产公司开发的其他公寓都采用同样的命名方式。如毕卡第公寓（今衡山饭店）对应皮卡第大区，培文公寓（今上海市妇女用品商店）对应旧省名贝阿恩，赛华公寓（今瑞华公寓）指代萨伏伊等。

诺曼底公寓之巍峨瞩目与其所处位置大有关系：基地本身为三角形，又处于五条马路交汇的岔道口，视野开阔，30度锐角头部倒成圆角比例仍显高瘦，昂首的气势颇似战舰。

这座八层高的大楼是沪上最早的外廊式公寓。平面因地制宜布置成熨斗形，北面沿福开森路设置2个开口改善采光通风。底层为商铺，沿霞飞路设置拱廊式骑楼。居室大多朝南，户型结构灵活，有1~4室户之分，共有63套公寓和30多间佣人房。内设电梯三部，消防楼梯多处，其中主入口门厅两部电梯采用指针显示楼层。

大楼外观采用法国文艺复兴风格，立面横向分三段：一二层基座为斩假石仿石墙面，中段三至七层用清水红砖，顶层檐部材质与基座相同，贯通的阳台和女儿墙构成双重水平线脚的檐部。底层半圆形拱券上设券心石，形成腰线的阳台采用宝瓶式石栏杆，三层和七层窗楣有山花装饰。

作为旧上海第一批高档公寓，该大楼最初住户多为外资公司高级管理人员。1946年后，万国储蓄会名下的楼盘多为政府收购，用作办公或高级官员宿舍，时任上海工务局局长的赵祖康（新中国成立前夕上海市代市长）曾分到诺曼底公寓的一套住宅。后来，著名文艺界人士如赵丹、郑君里、吴茵、秦怡等也曾在此寓居。

1943年，福开森路更名为武康路，1953年，诺曼底公寓更名为武康大楼。

“文革”期间，武康大楼有一个如今已被人淡忘的别名，叫“上海跳水池”。作为西区屈指可数的高层建筑，为数不少的无辜人士，不堪折磨在此跳下，令人扼腕。

1995年，武康大楼被列入上海优秀历史建筑保护名录。

参观指南

大楼外观和门厅可供参观，室内不对外开放。

I.S.S Apartments on Avenue Joffre (also called Normandie Apartments or the East Metro Apartments on today's Huaihai Road) was a signature Hudec work during his time with the R. A. Curry firm.

The building's unusual name led to wild speculative guesses. Many people had assumed that the name had something to do with the D-Day Normandie battle in 1944 during the World War II, but that seems rather unlikely as the building came up much before that.

The truth was closer to home. Fonciere et Immobillere de Chine, the real estate developer owned by the International Savings Society had named all their apartment buildings with names from French regions, such as Picardie Apartments (today's Hengshan Hotel), Beam Apartments (Shanghai Women's Department

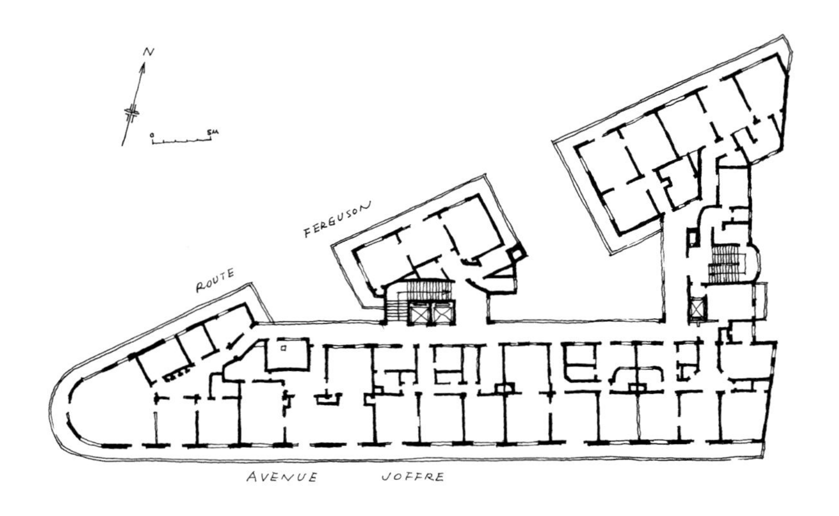

标准层平面 Typical Floor Plan

Store) and Savoy Apartments (Ruihua Apartments).

Normandie Apartments stands at a striking position. The triangle lot is located at the crossroads of five streets and provides an open view. The steep, sharp-headed building symbolizes a powerful warship.

The eight-story building was the earliest veranda-style apartments in Shanghai. The layout is shaped like a huge iron according to its lot.

Hudec had designed two entrances on its northern façade on the Route Ferguson to let in more light and air into the rooms.

The ground floor featured a rainbow of shops. The façade on Avenue Joffre is graced by arcade.

Most suites face south and are designed in a flexible layout ranging from one to four bedrooms. The building contains 63 suites and more than 30 rooms for servants. It is equipped with three elevators and several fire staircases. The two elevators at the entrance hall still show the floor number with pointers.

The building is designed in a French Renaissance style. The classic elevation is divided in three sections horizontally.

The external walls on the ground and second floors are coated with artificial stones while the floors above boast red bricks. Ashlar surfaced top floor correspondent with the base. The cantilevering balconies and the parapets juxtapose to become eave with double parallel corners. The three sections of continuous balconies form the vivid waist line on the façade.

Other noteworthy architectural details include the carved stones in every segmental arch of the ground floor, the vase-shaped stone balusters on the balconies and the classical pediments on the window lintels.

As one of the city's earliest high-end apartments, the building was initially leased to high-level employees of foreign companies. The Kuomintang government had purchased properties of the International Savings Society after 1946. Senior Kuomintang official Zhao Zukang, the substitute mayor of Shanghai before October 1949, had been assigned a suite in the building as his residence. Famous Chinese artists including actor Zhao Dan and actress Qing Yi also lived here.

Route Ferguson was renamed Wukang Road in 1943 and the Normandie Apartments became Wukang Building in 1953.

During the Cultural Revolution (1966-76), it earned the unfortunate moniker "The Shanghai Diving Pool". As one of the few highrise buildings in western Shanghai, quite a few people chose this building to end their lives by jumping from this classic Hudec work.

In 1995, the Wukang Building was listed a Shanghai Excellent Historical Building for Preservation.

Tips

The appearance and the entrance hall can be appreciated. The interior of the suites are not open to the public.

邬达克自宅 Hudec's Residence

记录建筑师天伦之乐的花园
A Garden of Happiness

邬达克在上海有三处自宅。哥伦比亚路 57 号(今番禺路 129 号)是第二处。

刚到上海在克利洋行工作时，邬达克租住在赫德路（今常德路)。1922 年 6 月 1 日，跟德国富商的女儿吉塞拉·梅尔完婚后，他开始设计第一座自宅。4000 平方米的基地位于吕西纳路（今利西路）17 号，靠近他当时正在建造的中西女塾。邬达克精心绘制了建筑和家具的设计图，并在岳父的资助下建造起来。

因为长子马丁、次子西奥多和女儿阿莱莎于 1923、1925 和 1928 年相继出世，邬达克决定另建更大的寓所，地点选在当时正在参与设计的哥伦比亚住宅圈附近。未料快建成时，这座西班牙式住宅转让给了孙科，于是他在哥伦比亚路对面重新购地，1930 年 6 月完成设计，5 个半月后新居建成。

这座自宅是典型的英国乡村风格，或称都铎复兴风格，最主要的特征是：露明木结构（并非真正结构，仅作为立面装饰)、陡峭的坡屋顶（上铺黏土红瓦，仿石板瓦效果)、屋顶老虎窗、立面直棂窗和高耸的烟囱等。这一形式或许显示了邬达克妻子的文化偏好，因为吉塞拉的母亲来自一个英国贵族家庭。

另一种可能同样不能排除：这是普益地产公司的项目，邬达克买下该基地时已完成部分结构。因为根据历史资料，建筑师买下的是曾规划了 6 座住宅的四块基地。而新设计主要集中在立面、室内装修、门窗、细部构造和增加的附房上。设计师本来最拿手的平面在这个项目中却没有太多研究，反而是屋顶和阁楼有细致的规划。更特别的是，这座住宅所用的建材品种规格不一，很有可能是邬达克收集了该地区其他在建项目的剩余物资，巧加利用的结果。

这座住宅南面拥有很大的花园，三个子女在此度过了整个童年。

然而，因系越界筑路，哥伦比亚路的市政管道一直存在问题。为了增铺管道和解决花园雨水倒灌等问题，从 1930 年 7 月起整整六年，邬达克不得不再三写信致电，与有关部门沟通，花费了不少时间和金钱，却始终未能妥善解决，令他头痛不已。

1936 年，邬达克将这座自宅转卖给一位德国外交官，和家人搬到自己开发设计的现代风格的达华公寓底层居住，直到 1947 年离开上海。

1964 年后，该住宅产权转给长宁区教育局。原来的花园上建起了旅游职校，住宅改为办公楼。现建筑经过改造更新，底层设邬达克纪念室。

参观指南

邬达克纪念室平日周二 13:30-16:00，
周日 9:30-12:00、13:30-16:00 向公众开放。

The fanciful house designed by Hudec as his own home is hidden from the often-jammed Panyu Road and behind an Italian restaurant and a white boutique hotel.

This Tudorbethan villa and its untended garden comprise a small wonderland — quiet and secluded from the city's hustle and bustle just some few meters away.

Hudec had three residences in Shanghai and the three-story house on 57 Columbia Road (now 129 Panyu Road), built in 1930, was the second one.

A newcomer to Shanghai, Hudec had rented an apartment on Hart Road (now Changde Road) when he worked for the R. A. Curry firm.

The young architect started to build his first

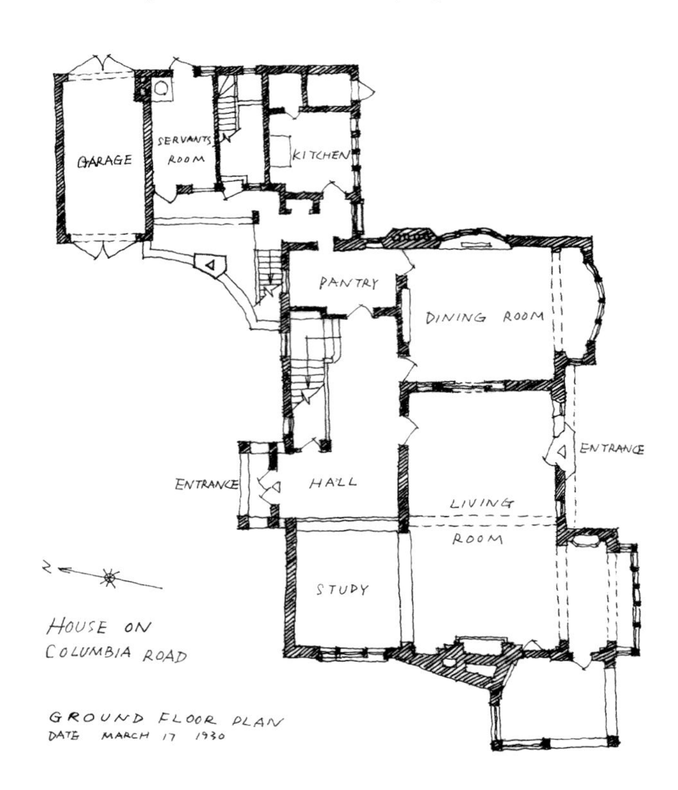

首层平面 Ground Floor Plan

residence after he got married to Gisela Meyer, the daughter of a wealthy German merchant, on June 1, 1922.

The 4000-square-meter site was located on 17 Lucerne Road (now Lixi Road), which was close to the McTyeire School, another Hudec work which was under construction at that time. Hudec designed both the building and the furniture himself. His father-in-law gave him financial support.

As his three children arrived quickly between 1923 to 1928, Hudec decided to build a larger home close to the Columbia Circle, a property he served as the architect. He designed a Spanish-style villa across the street for his family but later sold it to Sun Ke, son of Dr Sun Yat-sen.

Despite the financial crisis across the globe in 1929, Shanghai's real estate industry was relatively unaffected and there was in fact a building boom. Columbia Circle was an up-scale residential area for foreigners seeking respite from the hectic city and a bit of tranquility. Here, he designed villas in a rainbow of styles, including Dutch, Italian, Spanish and even several American styles.

But for his own home, he chose the quite traditional Tudorbethan style, probably according to the preference of his German wife Gisela, whose mother was from a British aristocratic family.

There was another possibility that the house was already built in half when Hudec bought the site. Smart Hudec had re-designed the façades, doors, windows and the interior without changing the layout too much. Moreover the residence was built with a variety of materials, which likely came from several projects he was building.

The Hudec House seems a pretty nice home — comfortable, functional and, most importantly, carrying an ambience of idyllic country life.

This is a white-wall building highlighted by contrasting dark-wood half timbering frames.

Two towering red-brick chimneys stand at each end of the sloping roof that is covered by beautiful slate tiles. Three sets of Gothic windows and a round arched door are on the ground floor. A big garden lies to the south.

The interior decorations were also the original creations of Hudec. Exquisite cabinets, lamp shades, parquetry wooden floors, western furniture and molded ceilings created a romantic and warm ambience.

The exterior is well preserved but the original furnishings inside are gone. Only the white molded ceilings in lovely patterns have been maintained.

Hudec and his family lived in the villa from 1931 to 1936. But the family later moved to another Hudec building, Hubertus Court (today known as the Da Hua Hotel) on West Yan'an Road because of poor drainage that made the house very humid and unhealthy.

After 1949, the house was used as a middle school and modern buildings mushroomed around this country villa. The ground floor of the house has been renovated to the Hudec Memorial Hall.

Tips

Hudec Memorial Hall is open from 1:30pm to 4:00pm on Tuesdays, from 9:30am to 12:00am, from 1:30pm to 4:00pm on Sundays.

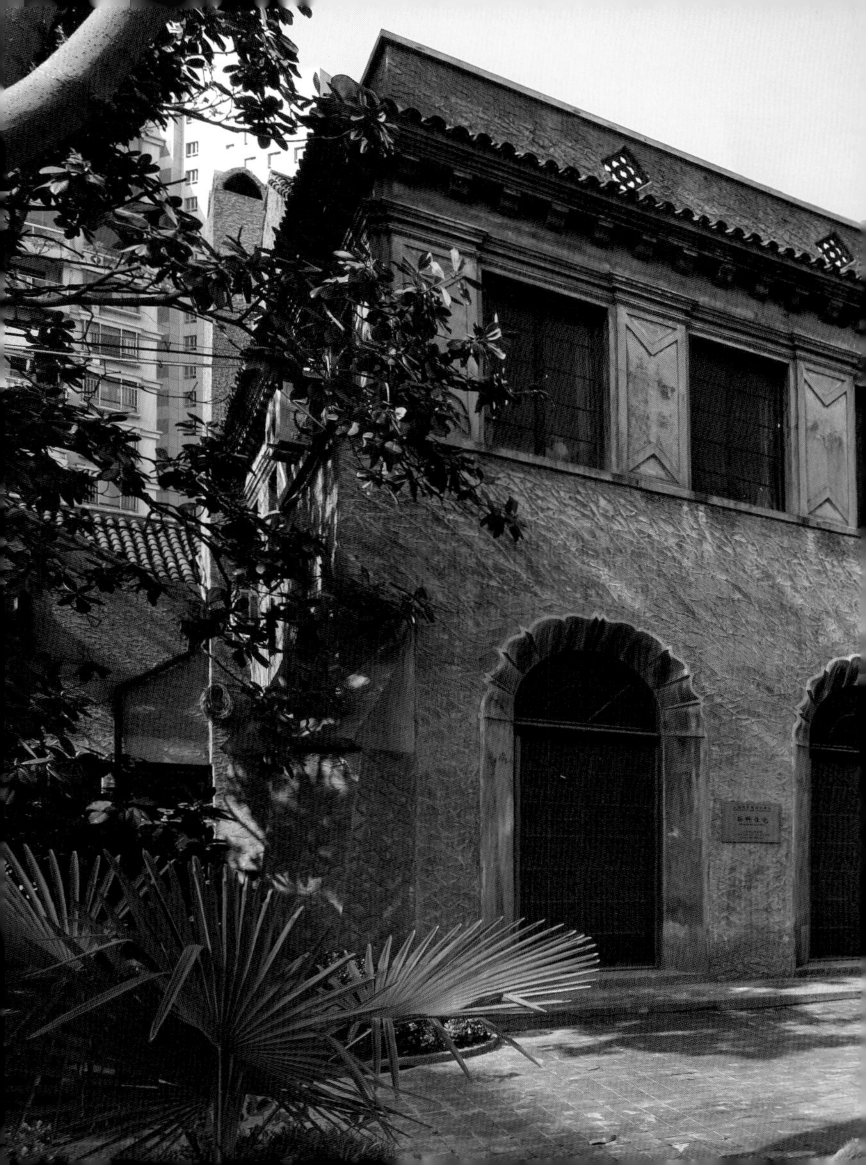

孙科住宅 Sun Ke's Residence

因感恩而低价转让的西班牙风格住宅
Hudec's Special Deal for Sun

孙科是孙中山先生的独子，系与原配夫人卢慕贞所生。早年留学美国，27岁后步入政坛，曾担任过国民党五院中三院的院长，财政部、建设部、交通部、铁道部部长，广州市市长和上海交通大学校长。1932－1948年，孙科长期工作生活在南京，回上海时就住哥伦比亚路22号（今番禺路60号）的住宅,陈淑芬、蓝妮两位夫人及其女儿也曾在此寓居。

孙科住宅占地约8000平方米，建筑面积1000平方米左右，砖木混合结构，假三层。这栋花园洋房本来是邬达克为自己兴建的，因设计慕尔堂时遇到麻烦，孙科曾帮忙解决，出于感恩低价转让给孙科，自己在哥伦比亚路对面另建新居。

该住宅具有明显的西班牙建筑特征，包括：平缓的红色筒瓦坡屋面、出檐很小、檐下半圆齿形装饰、尖券门廊、正立面窗间及室内的螺旋柱、烟囱的小尖券、土黄色鱼鳞状的拉毛粉刷等，同时融合了其他多种风格的元素。比如南立面二层中间用塔司干柱式支撑的圆拱券、窗间用几何图案装饰的浅壁柱和部分壁炉装饰属于意大利文艺复兴风格。底层落地窗周边的贝壳形拱券则有巴洛克的味道。

室内装修考究，楼梯全部用柚木，主要房间地板用柳桉木镶嵌成席纹，并都装有壁炉，且装饰各异。底层书房旁八边形的会客厅最有特色，地坪有一级高差，天花形成交叉拱，每个墙面开一扇尖券窗，与书房相连的入口部位则由两根矮胖的螺旋石柱托起一个尖券，深色的地面、柱式、窗框与白色的墙面和天花形成强烈的对比。

新中国成立前夕，孙科夫妇前往香港与母亲同住，后又移居美国。

此后，孙科住宅成为上海生物制品研究所的办公楼，大门移到延安西路1262号。因为涉及国家机密不对外开放，迄今仍显得颇为神秘。

参观指南

建筑不对外开放，从番禺路透过围墙可以窥见部分外观。

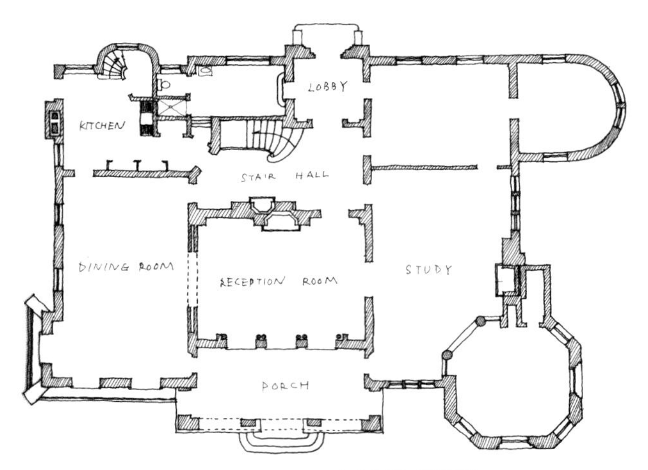

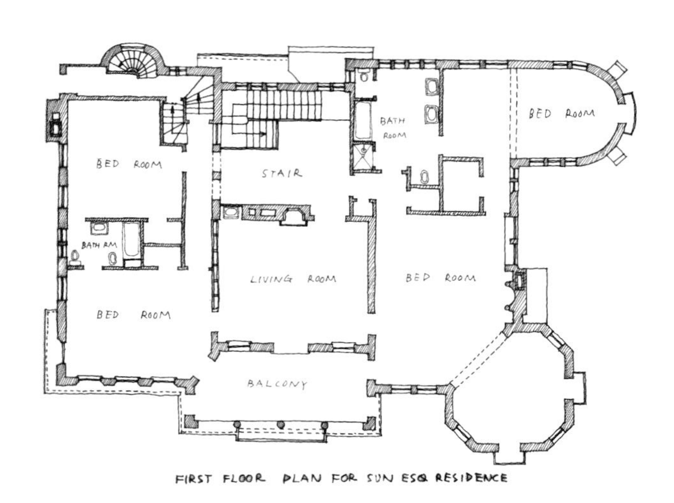

首层、二层平面 Ground Floor Plan+ 1st Floor Plan

The beautiful former residence of Sun Ke, the only son of Dr. Sun Yat-sen (the founder of Kuomintang), is a hard-to-find treasure.

The three-story cream-painted villa at 60 Panyu Road sits amidst the lush gardens of the state-owned Shanghai Biochemical Research Institute which has used the house as its office since the 1950s.

Hudec built the house (whose entrance is now on West Yan'an Road) in the late 1920s as his own residence. But he sold the house to Sun Ke at a low price because of a favor Sun had done for him.

Hudec's innovative architectural skills can be seen everywhere in the 1,000-square-meter, brick-and-wood construction.

The overall style is typical Spanish, from its sloping roof topped by red mission tiles, pointed-arch atrium, spiral columns and yellow stucco shaped like scales on the wall.

The design is cleverly juxtaposed with other architectural elements. Italian renaissance style can be found on the arch supported by Tuscan orders on the south façade, on the shallow pilasters with geometrical patterns and on some decorations of fireplaces. The shell-shaped arches showcase a Baroque style. Columns, in a range of shapes and patterns perfectly grace the study on the first floor and the hallway on the second floor.

It seems that he designed the house in a free, creative mood. Perhaps because the house was originally meant for himself, he had the freedom for creation.

The staircase is paved with teakwood while the parquet floor is made of clauan wood. Every major room is equipped with a fireplace in a different style. The centerpiece of the first floor is an octagonal sitting room next to the study.

Born in 1891 in South China's Guangdong Province, Sun Ke was the son of Dr. Sun Yat-sen and his first wife, Lu Muzhen. Dr. Sun divorced Lu in 1913 to marry Soong Ching Ling, the second of the three famous Soong sisters.

Educated at the University of California and Columbia University in New York, the younger Sun had been the mayor of Guangzhou, capital of Guangdong Province, for three terms between 1921 and 1926 and proved to be a good public official. He improved the education system, the environment and public security.

Sun Ke lived in the house on Panyu Road from 1929. He had to spend most of the week in Nanjing where he was an official with the then Kuomintang government. However, he was able to get back to Shanghai on most weekends.

Sun moved to the US in 1952 and died of a heart attack in Taipei on September 13, 1973.

The building is not open to the public. The inaccessibility adds to a sense of mystery to the hard-to-find architectural gem created by a free-wheeling architectural genius.

Tips

The building is not open to the public but part of its appearance can be appreciated through the walls on the Panyu Road.

交通大学工程馆
Engineering and Laboratory Building of Chiao Tung University

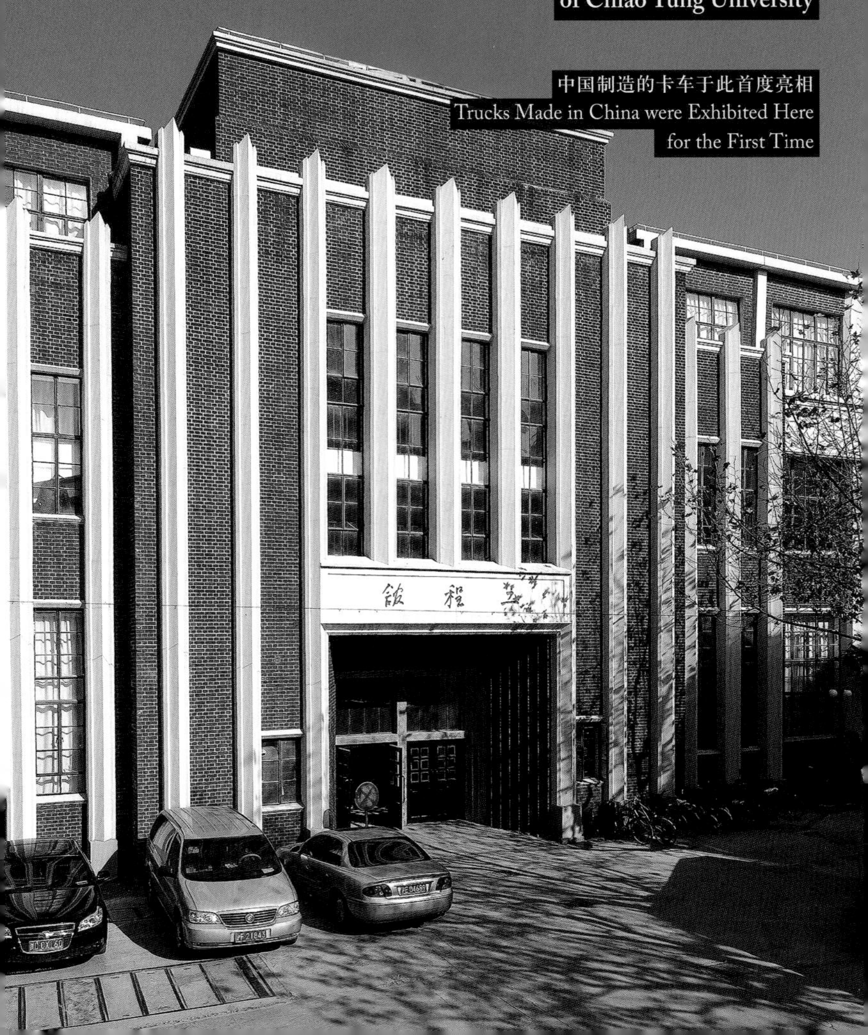

中国制造的卡车于此首度亮相
Trucks Made in China were Exhibited Here for the First Time

交通大学工程馆是邬达克的作品，了解这一情况的人还不算少。然而鲜为人知的是，今天上海交通大学徐汇校区的基本格局也与邬达克相关。

著名实业家、洋务派领袖之一盛宣怀 1898 年创办的南洋公学，于 1921 年改为交通大学。到 1929 年前后，学校借政府加强高等教育投资的契机，拟扩建校园。时任铁道部长兼交通大学校长的孙科，因为邬达克有转让自宅的交情，就委托后者对校园扩建工程进行整体规划。包含执信西斋（范文照设计，1929 – 1930)、铁木工厂在内的一批新楼在其后两年相继建成。

当然，邬达克在此最重要的设计还是工程馆。这座于 1931 年 2 月由孙科奠基的实验、教学、办公综合大楼是由张元济、王清穆、唐文治、蔡元培、陆梦雄等老校长会同部分老校友发起，并募集资金建造的。

这是一栋回字形的建筑，中间有很大的内院。底层设有锅炉房、机械、水力、金工、材料、电气、标本等各种试验室，上层设有教室、绘图室、仪器室、阅览室、教授休息室、摄影暗房等，还有两个大报告厅。建筑原高两层，在 1960 年增建至三层，建筑面积由 1.1 万平方米扩大至近 1.3 万平方米。

大楼采用钢筋混凝土结构，外观简洁，属于现代装饰艺术风格。外墙大部分采用深褐色泰山面砖，石材壁柱凸出墙面呈锯齿形，强调竖线条，其立面表现手法与广学、真光大楼类似。南立面入口叠涩的门洞形成浓厚的阴影，朝向内院的北立面入口则是三个尖券门，透露出邬达克对哥特式风格的特别偏爱。门厅内大楼梯分设两侧，几乎占据整个空间。教室区域走廊有三米多宽，一侧还沿墙面设水磨石长座凳，充满宁静的学习氛围。该楼同样是邬达克与老搭档馥记营造厂合作的结果。

事实上，大楼建造中间还有一段小插曲。1932 年 8 月建成的工程馆首先被用作了工业展览馆，因为 1933 年春，交通大学要举办第三届工程与交通展览会，中国制造的卡车将历史性地首次亮相。

1947 年，为纪念老校长叶恭绰的功绩，工程馆曾更名为“恭绰馆”，1949 年后恢复原名。在这座以机械、电机工程为重点的大型教学楼里，培养出了钱学森（1934 届机械系）、江泽民（1947 届电机系）等卓越人才。

参观指南

上海交通大学徐汇校区校园内的教学建筑，基本开放。

Very few people are aware that Hudec had designed a building inside the Shanghai Jiao Tong University. In fact, he had even made a plan for the Xuhui Campus of the university.

The Normal College of Nan Yang Public School was founded in 1898 by a Qing Dynasty (1644-1911) official Sheng Xuanhuai, who was a promoter, shareholder and manager in nearly all the official business ventures set up in Shanghai between 1871 to 1895.

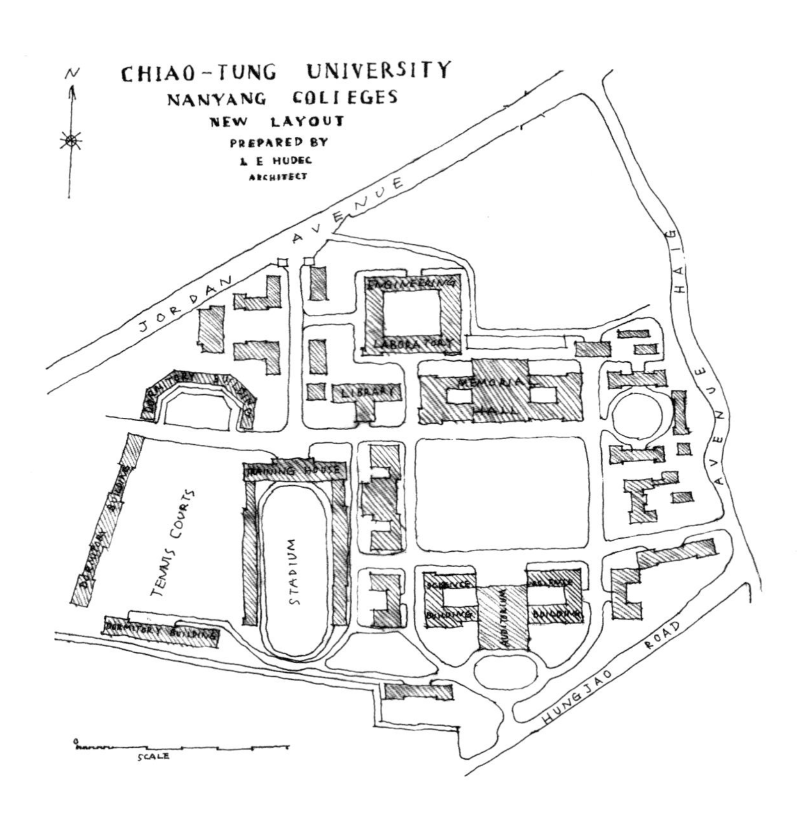

规划总平面 Layout for Campus

The college became Chiao Tung University (today's Jiao Tong University) in 1921 with plans to expand the campus in 1929 with government support. Dr. Sun Yat-sen's son Sun Ke, who was the minister of railways and the head of the university, commissioned Hudec to draft an overall plan for the new campus. Hudec had done him a favour before by selling him a residence at a low price. Within two years a group of new buildings, including a new students' residential building named Zhi Xin Xi Zhai (designed by Fan Wenzhao 1929-1930) came up.

The Engineering and Laboratory Building designed by Hudec was built with funds raised by several former university leaders and graduates including Zhang Yuanji and Cai Yuanpei. In February 1931, Sun Ke laid a corner stone for the building which was intended for research and experimenting, teaching and administration.

The building is enclosed by walls and has a large yard inside. The first floor featured a boiler room and labs for studies like mechanics, hydraulics, metalwork, material science, electrics and specimen. Classrooms, drawing rooms, reading rooms, professors' restrooms, apparatus room, and a darkroom for photography as well as two lecture halls are all on the second floor.

Another floor was added to the two-story building during renovation in 1960, which increased the building area from 11,000 to 13,000 square meters.

The steel-and-concrete lab building has a simple-cut façade in Art Deco style. The external walls are coated with dark brown tiles. The edged stone pilasters stretch beyond the walls, emphasizing the vertical lines. The expression technique of the façade is similar to the one at the Christian Literature Society Building and the Baptist Building.

The cantilevering doorway on the south forms a thick shadow. The three pointed arches on the north entrance reveals Hudec's preference for Gothic style.

Perched at the two sides of the hall are the two grand staircases which dominate almost all the space. The corridors are as wide as three meters and are designed with terrazzo benches on one side along the wall, which exudes a serene, quiet atmosphere for studying.

When the building was completed in August 1932, it was first used as an industrial exhibition hall since the university was about to host the 3rd Exposition for Engineer and Communications in the spring of 1933, during which trucks made in China were exhibited for the first time.

The building was renamed Gongchuo Building in 1947 in memory of the former university head Ye Gongchuo but reverted to its original name after 1949. A galaxy of Chinese intellectuals graduated from this well-designed Art Deco building including famous Chinese Scientist Qian Xuesen, who graduated in mechanics major in 1934, and former Chinese President Jiang Zemin, who graduated in electrical engineering from here in 1947.

Tips

Like other buildings in the Xuhui Campus of Shanghai Jiao Tong University, it is open to the public.

哥伦比亚住宅圈 Columbia Circle

西郊宣传声势最大的花园住宅项目
The Most Advertised Garden Villa Project in Shanghai's Western Suburbs

民国十三年（1924）9月，江浙两地军阀在上海周围恶战，租界当局乘机越界筑路。次年，工部局在沪西建成13条新路，哥伦比亚路（今番禺路）和安和寺路（今新华路）也在其中。

由于当时广泛的政治运动，很多外国人都深感威胁，美商雷文创建的上海普益地产公司敏锐地觉察到商机，在这一安静的城郊区域购地百余亩兴建高档住宅区，宣传语还特别强调“经美领馆注册，并有外国事务委员会盖章的外国所有权”，可“免于讨厌的入境骚扰”。

当时统一规划的用地共有82块，安和寺路以北为沿街基地，以南则腹地较深，除Y形主通道外，还有4条支弄垂直城市道路。每块基地拟建造不同风格的花园洋房，整体命名为“哥伦比亚住宅圈”或“普益模范村”，宣传书往往把北面不远处的哥伦比亚乡村俱乐部（即邬达克设计的美国乡村总会）也纳入区域范围。

该地产公司的执行经理山多尔是匈牙利人，他聘请同胞邬达克进行建筑设计。这些二层或三层的独立式住宅只有5个标准户型，但外观却有13种样式，5种英国式，6种美国式，还有意大利式和西班牙式，当然，这些建筑风格大多是混合的，建筑师还可以根据客户需要增减房间。普益地产同期请邬达克在西爱咸斯路（今永嘉路）、巨福路（今乌鲁木齐南路）、亚田南路（今安亭路）设计的多幢别墅，也采用这些通用户型和立面。

哥伦比亚住宅圈建于旧上海花园住宅开发的顶峰期，不过这种风潮很快过去了，到1932年，真正动工的地皮约40块,邬达克至少设计了其中26个项目。

从当初地块规划图来看，哥伦比亚住宅圈不仅包括现在长宁区新华路211弄的外国弄堂和329弄的新华别墅，也包括现新华路以南的119弄、155弄、185弄和以北从番禺路到香花桥路之间的沿街地块。留存下来，当年独门独户的洋房现在通常住着6~8户人家，树木茂盛的花园里种着果蔬，晒着衣被，还有的成了幼儿园或酒吧，是另一派都市图景。

需要指出的是，邬达克在此设计的主要是标准化的房产项目，一些风格和形式比较独特的建筑，包括329弄36号国内唯一的圆形别墅——“蛋糕房”是否出自他的手笔还有待进一步考证。

参观指南

该区域现多为私宅，内部不对外开放，沿新华路或在弄堂内可以欣赏外观和花园。

14

Following fierce battles near Shanghai between warlords from Jiangsu and Zhejiang Provinces in 1924, the Municipal Council set about constructing 13 "Extra-Settlement" roads. By building roads outside the concessions, they further expanded the territory and even sent policemen to manage the roads after they were completed. Columbia Road (now Panyu Road) and Amherst Avenue (now Xinhua Road) were among the Extra-Settlement roads.

With a nose for business, American merchant Frank Raven and his Asia Realty Company bought vast land area of as much as 100 mu (around 66 667 square meters) in the extra-settlement neighborhood for high-end property, which was still suburban land then.

The site contained 82 pieces of lots located at the south of the Amherst Avenue, including a Y-shaped main road and four subsidiary roads. Each piece of lot was planned for a garden villa of different styles. The property named Columbia Circle also included the American Country Club.

Project manager Hugo Sandor was a Hungarian, who employed his compatriot Hudec as the architect. The two or three-story villas had only five standard layouts. However, a customer could choose from a rich selection of appearances in 13 styles, including an Italian, Spanish, five British and six American styles. The architect also promised to customize the number of rooms according to the individual needs.

Asia Realty Company had also invited Hudec to design several villas on Route Sieyes (now Yongjia Road) and Rue Adina (now Anting Road) with the same layouts and façades.

Columbia Circle was built at the peak of garden villa vogue in Shanghai, but which soon faded away. By the year 1932 only 40 pieces of lots were under construction, 26 of which were designed by Hudec.

According to the original layout plan, Columbia Circle included not only the famous "foreign lane" on 211 Xinhua Road and the Xinhua Villa on 329 Xinhua Road, but also lanes on No. 119 and No. 155 Xinhua Road. The land between Panyu and Xianghuaqiao Roads also belonged to the property.

Today the surviving villas are shared by six to eight families. The lush gardens are planted with fruits, vegetables and dotted with air-drying clothes and quilts. Some have been renovated into a kindergarten or a bar, showcasing vivid urban life.

Hudec had designed the villas here in standard forms. It is still not known if some strange-shaped buildings in the neighborhood, such as the striking round-shaped "Cake Villa" were actually Hudec's work.

Tips

These are private residences and are not open to the public. The exteriors and the gardens can be appreciated from Xinhua Road or nearby lanes.

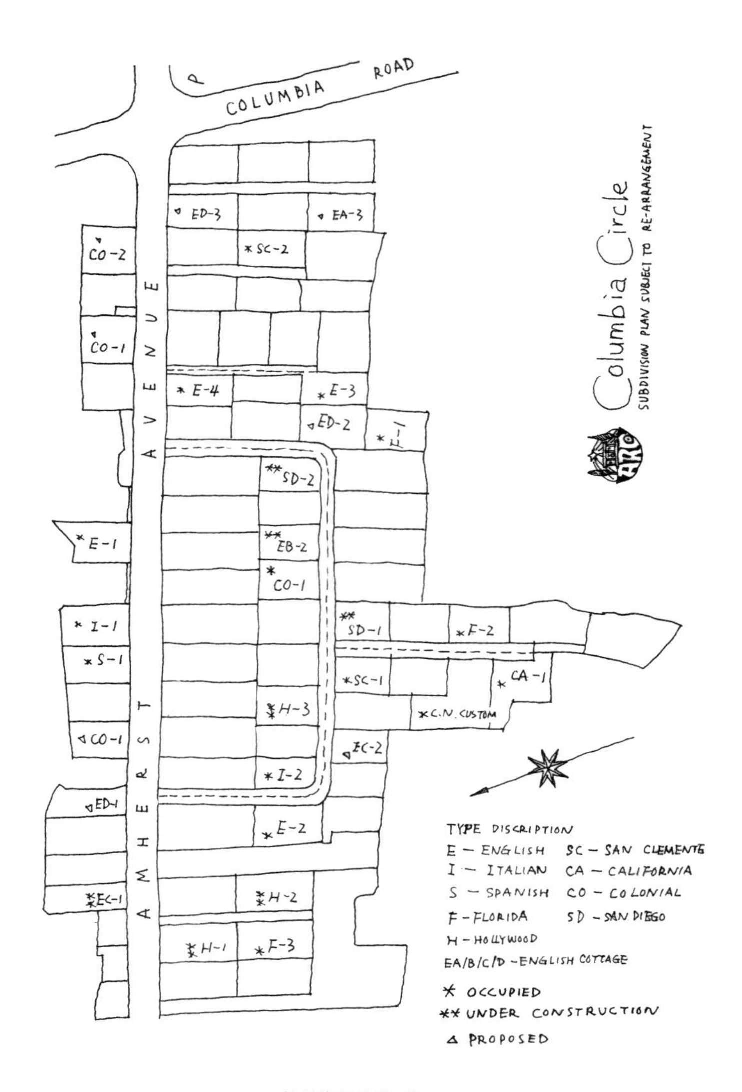

规划总平面 Site Plan

达华公寓 Hubertus Court

邬达克自行投资兴建的公寓
An Apartment Invested by Hudec Himself

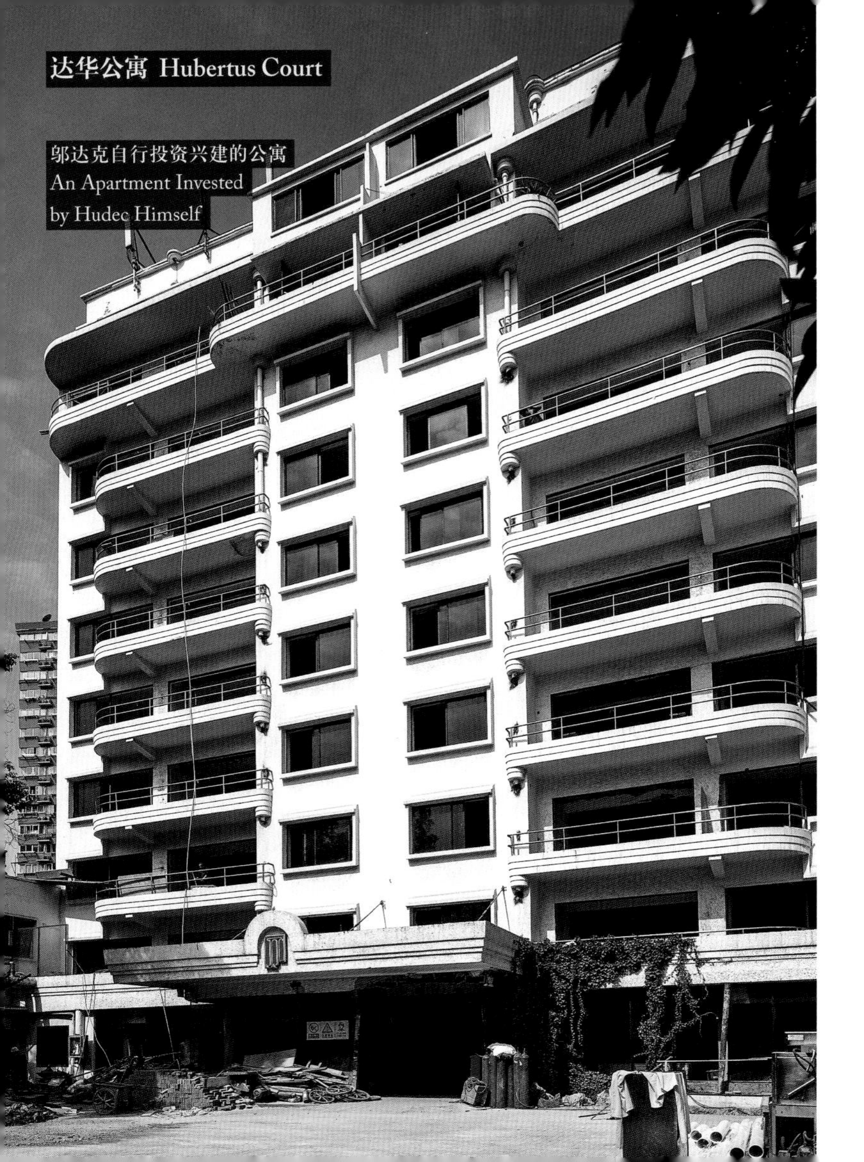

20世纪二三十年代是近代上海经济最为繁荣的阶段，也是房地产业最为鼎盛的时期。房地产的暴利吸引越来越多的人投身该行业，邬达克也不例外。同时，由于租界人口倍增，能建造花园住宅的基地越来越稀缺和昂贵，到1930年后，高层公寓取代办公楼和低层洋房，成为城市中最主要的建筑类型。

1935年，邬达克在美国领事馆注册了自己的开发公司——邬达德房产联合公司，名下的房产除哥伦比亚路自宅外，最主要的就是在大西路（今延安西路）的这幢十层公寓楼——达华公寓。

达华公寓地处英美租界西边界以外越界筑路的区域，离哥伦比亚路不远。该公寓是当时这一地区最高的建筑，周围多为华界传统坡屋顶的低矮住宅，又因设计现代，故有“小国际饭店”之称。据邬达克长子回忆，是吉塞拉的父亲承担了大楼的主要建设费用。

大楼底层北面沿街是门厅和电梯，南面是面对大花园的一套公寓，建筑师留作自用。最西一跨架空成为车道，毗邻的东面地块停车场共设20个车位。标准层一梯三户，两侧各布置一套大公寓，朝南设大阳台，北面是厨房和佣人房。为满足消防要求，北立面东西两角各有一个圆形疏散楼梯。中间一户是带两个房间的小公寓。楼顶两层平面变为带露台的两套公寓，立面上自然形成变化。

达华公寓的设计明显受到欧洲现代住宅的影响，体形简洁明快，水平线条流畅。西班牙和日本的专业杂志刊登了这一作品，并将之与英国以及布拉格和米兰的一批住宅建筑相提并论。在同期日本杂志中还包括纽特拉、格罗皮乌斯、阿尔托等那一时期的顶尖建筑师的作品介绍，这再次表明邬达克当时已具有国际知名度。

1937年，邬达克将两位儿子送到加拿大维多利亚一所军事院校学习后，家里只剩三个人。他将哥伦比亚路的房子出让给一位德国外交官，携妻女搬到达华公寓的底层公寓居住，直至1947年初夏全家离沪。

1949年后，达华公寓先后改为达华饭店、达华宾馆，并新建三层南楼，增建四层东楼。

1999年，该楼被列入上海市优秀历史建筑保护名录。经过一年多的改造，2013年底，达华公寓以设计酒店的形象全新面世。

参观指南

沿延安西路或进入南面花园都可以欣赏建筑外观，在延安路高架上景观尤佳。大厅等公共部位也开放，客房区域不开放。

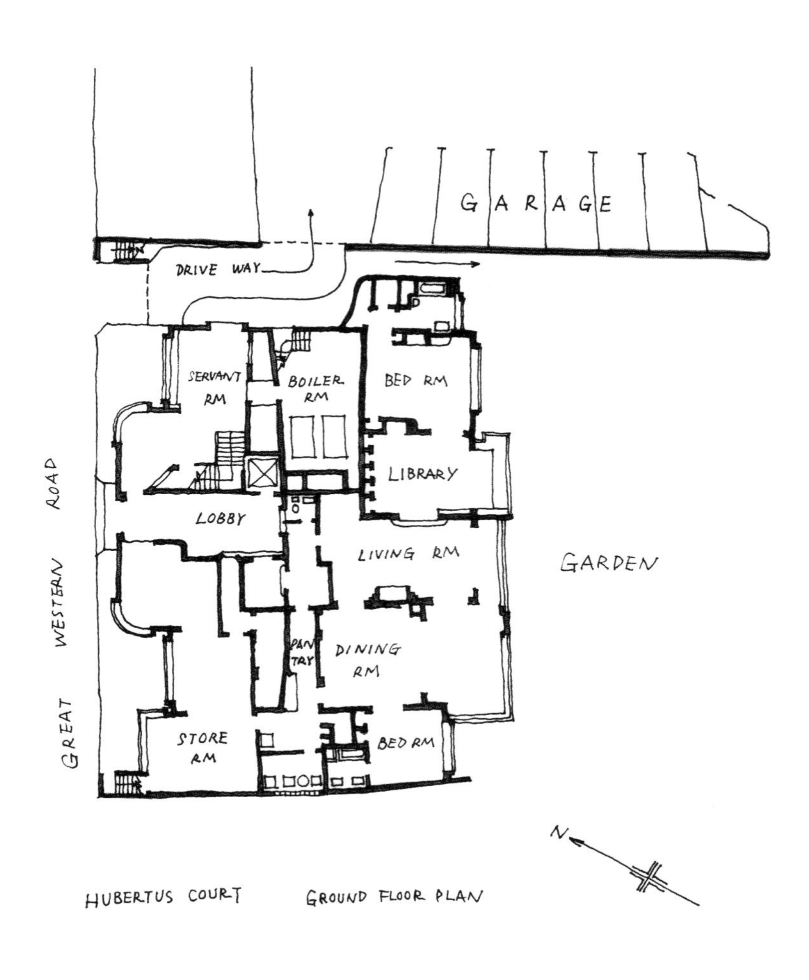

首层平面 Ground Floor Plan

Nicknamed the "Little Park Hotel," Hubertus Court (now the Da Hua Hotel) was designed, invested and dwelled by Hudec who spent his last ten years in Shanghai in the ten-story apartment building.

Shanghai had enjoyed a golden period in the 1920s and 1930s which understandably led to a construction boom in the city. Huge profit had attracted hordes to the city who invested in real estate, including Hudec. Due to the growing population and decreasing land, tall apartment buildings became the norm instead of office buildings and garden villas after the 1930s.

Hudec registered his own developing company, the Hubertus Properties Fed. Inc at American Consulate Shanghai. Hubertus Court on Great Western Road (now West Yan'an Road) was the main project of his company.

Located at the extra-settlement region outside the western edge of the International Settlement, Hubertus Court used to be the tallest building among a group of traditional Chinese buildings in the region. The modern style had earned a nickname, "the Little Park Hotel." According to Hudec's eldest son, the construction for the building was financed by Hudec's father-in-law.

The entrance was from the ground floor, which also had an elevator and a suite that stared out to the big garden which the architect used it for himself. The parking lot had space for 20 vehicles. A standard floor included three suites. Two spacious suites were arranged at the two sides with a big balcony facing south and a kitchen and servant's room in the north. A two-bedroom suite of a smaller size sat in the middle.

Hudec designed two staircases for fire escape on the northern façade. The top two floors have a different layout, each containing two suites with terraces.

The simple shape and smooth horizontal lines of Hubertus Court reflect a strong influence of modern European residence. Some architecture magazines from Spain and Japan reported on the building and compared it with a group of residential buildings in England, Prague and Milan. Some world renowned architects including Richard Neutra, Walter Gropius and Alvar Aalto also appeared in the same issue of the Japanese magazine, which implied a global recognition of Hudec at that time.

Hudec's family in Shanghai included just three people after his two sons went to study in a military academy in Victoria, Canada, in 1937. He sold his former home on Columbia Road (now Panyu Road) to a German diplomat and moved to live in the suite on the ground floor of the Hubertus Court with his wife and daughter until they left Shanghai for good in the early summer of 1947.

Hubertus Court was renovated as Dahua Restaurant first and then as Dahua Hotel after 1949. A three-story building in the south and a four-story building in the east were added to the hotel afterwards.

The building was listed as Shanghai Excellent Historical Building for Preservation in 1999. It reopened as a boutique hotel at the end of 2013 after a year-long renovation.

Tips

The façade can be appreciated from the garden and on the West Yan'an Road. The view from Yan'an Elevated Highway is perfect. The lobby of the hotel is open to the public but the hotel rooms are for guests only.

E 区

4. 何东住宅 (1919-1920)
   今商务办公楼
   陕西北路 457 号 (近北京西路)
   静安区

18. 爱司公寓 (1926-1927)
    今瑞金大楼
    瑞金一路 148－150 号 (近淮海中路)
    黄浦区

29. 刘吉生住宅 (1926-1931)
    今上海市作家协会
    巨鹿路 675 号 (近陕西南路)
    静安区

41. 斜桥弄巨厦 (1931-1932)
    今上海市公惠医院
    石门一路 315 弄 6 号 (近南京西路)
    静安区

51. 震旦女子文理学院 (1937-1939)
    今向明高级中学震旦楼
    长乐路 141 号 (近瑞金一路)
    黄浦区

Zone E

4. Ho Tung's Residence (1919-1920)
   now "The Masion"Office Building,
   No. 457 North Shaanxi Rd (near West Beijing Rd), Jing'an District

18. Estrella Apartments (1926-1927)
    now Ruijin Building,
    No. 148-150 Ruijin 1. Rd (near Middle Huaihai Rd), Huangpu District

29. Liu Jisheng's Residence (1926-1931)
    now Shanghai Writers' Association,
    No. 675 Julu Rd (near South Shanxi Rd), Jing'an District

41. P.C.Woo's Residence (1931-1932)
    now Shanghai Gonghui Hospital,
    No. 6 Lane 315 Shimen 1. Rd (near West Nanjing Rd), Jing'an District

51. Aurora College for Women (1937-1939)
    now Xiangming Senior High School Aurora Building,
    No. 141 Changle Rd (near Ruijin 1. Rd), Huangpu District

新闸路 Xinzha Rd
本图为位置示意，与实际比例不符
illustration is not proportional to the actual scale
轨道交通及车站
Metro Stations
步行时间 (分钟)
05 Walking Time (in minutes)
4
15
北京西路 West Beijing Rd
西康路 Xikang Rd
石门二路 Shimen 2. Rd
2号线 南京西路
Line 2 West Nanjing Road
南京西路 West Nanjing Rd
吴江路 Wujiang Rd
06
41
静安区
Jing'an District
陕西北路North Shaanxi Rd
茂名北路 North Maoming Rd
石门一路 Shimen 1. Rd
威海路 Weihai Rd
延安中路 Middle Yan'an Rd

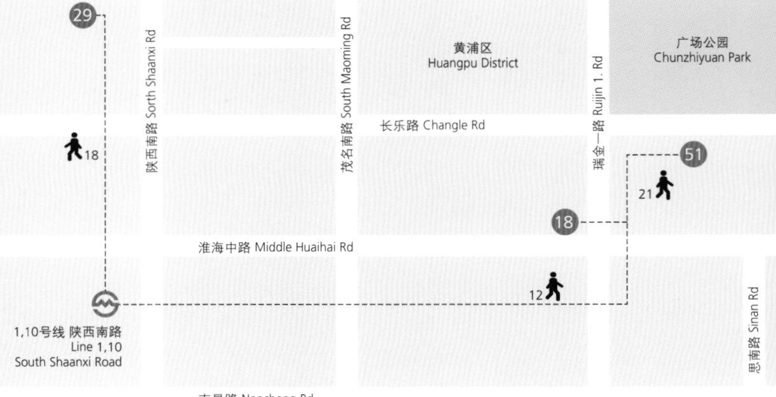
巨鹿路 Julu Rd
29
陕西南路 Sorth Shaanxi Rd
茂名南路 South Maoming Rd
黄浦区
Huangpu District
瑞金一路 Ruijin 1. Rd
广场公园
Chunzhiyuan Park
长乐路 Changle Rd
18
51
21
18
淮海中路 Middle Huaihai Rd
12
1,10号线 陕西南路
Line 1,10
South Shaanxi Road
思南路 Sinan Rd
南昌路 Nanchang Rd

何东住宅 Ho Tung's Residence

香港首富的古典豪宅
Classic-style Villa of Hong Kong's Most Wealthy Man

何东为香港开埠后的首富，近代上海房地产巨商，汇丰银行等多家企业和公司的主要股东，今澳门赌王何鸿燊是其侄孙。因成就显赫，曾获英、中、法、德、意大利、葡萄牙等多国的爵士或勋位。其父为荷兰裔犹太人何仕文，后入英国籍，母为祖籍广东的施氏。

何东住宅是邬达克初到克利洋行时在法租界为外国富商设计的多所古典风格豪宅之一，原业主为赫希斯特生物公司上海分部的总裁卡茨，何时转给何东则不得而知。

该建筑为二层混合结构，平屋顶，采用新古典主义风格，檐口厚重，主立面多以爱奥尼式列柱或壁柱予以强调。如起居室外南立面门廊有四根贯通两层的巨柱，西侧半圆形阳光房的外墙中间嵌有两根，东面主入口上方则是两组双壁柱。东、南两面二层的弧形阳台均由雕饰精美的牛腿支撑。

室内亦装饰华美。门厅为黑白相间的大理石铺地，印花壁纸墙面，房间铺柚木地板，饰雕花护壁，吊顶的石膏线脚也各不相同。主要房间均设有壁炉，且形式各异。壁炉与木家具的雕饰都是精工细作，部分陈设由邬达克亲自设计。锅炉房、暖气水汀等配套设施也一应俱全。

根据邬达克的家书，他在现场监理了该住宅的施工。当这座土建造价 4.5 万美元，又用 2 000 美元装饰的豪宅几近完工时，却不幸遭遇了一场大火灾，公司不得不从头开始，再次建造。

1949 年后，何东家族迁回香港。1958 年，中华书局《辞海》编辑所（今上海辞书出版社）进驻何东上海公馆。近年，该楼被改造为商务办公楼。

参观指南

在花园可以欣赏建筑立面，但室内不对外开放。

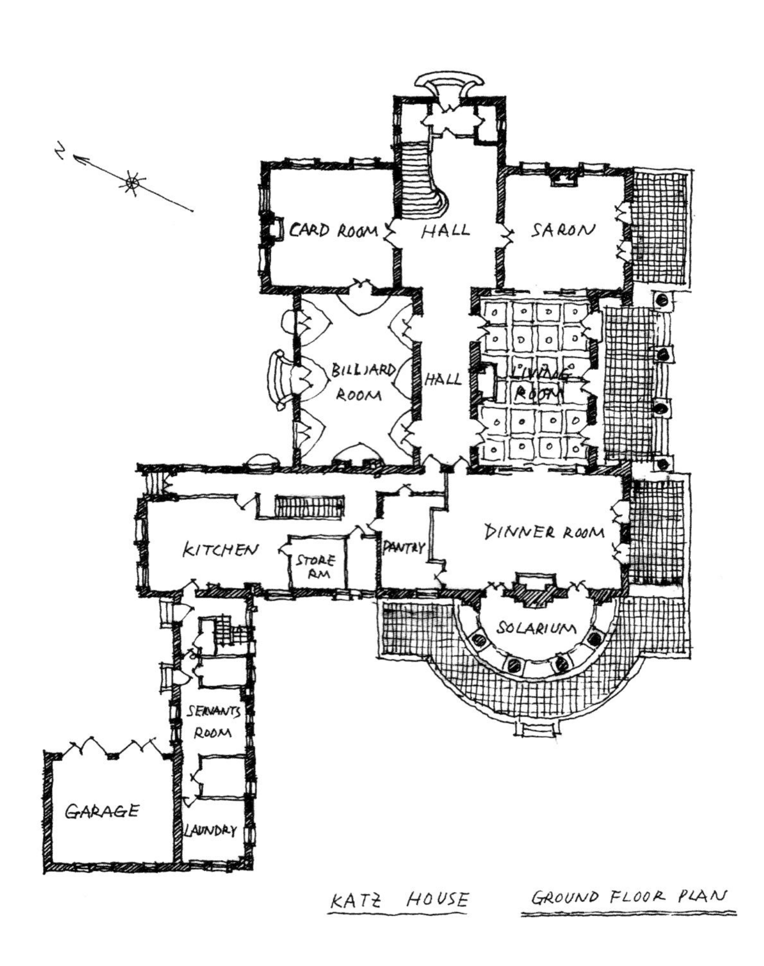

首层平面 Ground Floor Plan

The residence of Hong Kong tycoon Sir Robert Ho Tung mirrored another Hudec work — tycoon Liu Jisheng's residence on Julu Road .

Both villas are his early works and are essentially neoclassical in style. The lesser known Ho's villa on Shaanxi Road was initially designed for Wm Katz, director of the Shanghai Plant of Hoechst Chemicals, who later sold it to Ho Tung.

Sir Ho was born in 1862 in Hong Kong where he became the wealthiest man of his time and a major stockholder in many enterprises including the HSBC Bank.

Keen on public affairs and charity, he was knighted by the British Crown for his donation during World War I. And during the long war he also increased his investment in Shanghai and acquired quite a lot of properties in the north Bund area. Macao billionaire Stanley Ho Hung Sun, nicknamed "The King of Gambling", is his grand nephew.

The building's east gate is the first thing that impresses any visitor, which is graced by a huge lintel with triangular decoration and two Ionic columns that support a mini-balcony.

The front — the south façade — is in the same style but on a much grander scale. Four huge Ionic columns mushroom up to the second floor, which makes a nice porch and elegantly hides a shallow balcony on the second floor.

White French windows in interestingly different shapes and sizes, a flat roof surrounded with parapets and railings — everything you could imagine for a dream house, this has it all.

Inside, the building retains a strikingly beautiful black-and-white marble floor and the original grand spiral staircase with its patterned iron railings. The fireplaces and wood furniture were embellished with exquisite carvings.

According to Hudec's family letter, he was supervising the construction on the site. Unfortunately the villa which cost up to US$47,000 to build was destroyed in a big fire when it was near completion. They had to start the project all over again.

Ho's family returned to Hong Kong after 1949 and the state-owned Shanghai Lexicographical Publishing House used the villa after 1958. It was renovated and then leased to a company providing service offices mainly for foreign companies.

Tips

The façade can be admired from the garden but the interior is not open to the public.

爱司公寓 Estrella Apartments

霞飞路上的明星公寓
The "Star" Apartment
on Avenue Joffre

霞飞路对旧上海法租界而言，地位相当于公共租界的南京路，是最昂贵的区域，高档住宅、商店和外国领事馆在此云集。诺曼底公寓建成后，邬达克在霞飞路还设计过一座高层住宅——爱司公寓。

公寓名 "Estrella" 在西班牙语中有明星的意思。的确，在当时的霞飞路，无论从体量、风格还是颜色上来看，爱司公寓都可谓鹤立鸡群。

该项目的业主是 1910 年后从近东来沪的犹太裔商人科亨，最初通过经营黄包车生意致富，后与中国公司合作开发房产。

爱司公寓位于霞飞路和圣母院路（今瑞金一路）转角，是一座七层商住楼，共有 15 套独立公寓。底层为商铺，每个店面内有一座独立小楼梯，通往二层的两居室公寓。三层以上每层有两套三居室带佣人房的公寓，一大一小，分别朝向两条街道。两者间设楼梯、电梯和过厅，屋顶有花园。

当时媒体称其为沪上最早拥有如此现代化设施的公寓，内容包括：垃圾焚化炉（厨房外生活阳台区）、机械式冰箱、按季节分设的衣橱、内置式烫衣板等。功能设计也很周到：门厅内有放置衣帽、雨伞的柜子和照身镜；厨房配备煤气或电炉、现代化水槽；卫生间有弓形浴缸和淋浴房，镜子带照明以方便剃须，等等。

建筑外观追随西班牙地方风格，并融合文艺复兴和巴洛克的装饰元素。建筑立面横向分三段：一二层为基座，三至六层立面间隔设置凸肚窗和带铸铁花饰的浅阳台，顶部由连续的阳台和檐口构成。窗楣花饰、梯形凸窗中间竹节状的窗棂、室内公共部位地面及墙裙用马赛克镶嵌的几何图案，这些是较明显的西班牙建筑特征，但在外立面砖红色镶边、线脚与整体米黄色粉刷墙面的鲜明对比中，又能看到东欧传统建筑的影子。

当时，霞飞路中段被称为“圣彼得堡的涅瓦街”，因为从马斯南路（今思南路）到善忠路（今常熟路）之间多为精美的俄侨商店，散发着浓郁的斯拉夫气息。爱司公寓的异域风格正与之协调。

1943 年，圣母院路更名为瑞金一路，爱司公寓后改称瑞金大楼。1999 年大楼被列入上海市优秀历史建筑保护名录。

参观指南

沿淮海中路瑞金一路可欣赏外观，建筑室内不对外开放。

Avenue Joffre (now Middle Huaihai Road) was the most expensive community in the former French concession, similar to Nanking Road in the International Settlement. High-end residences, luxury shops and foreign consulates set up bases along the road where Hudec built the famous Estrella Apartments.

Estrella means "star" in Spanish which seems apt for this residence — a true "star" it was in terms of the size and the style.

The property owner was Jewish merchant Alberto Cohen, who had moved to Shanghai from the Middle East after 1910. He made a fortune from the rickshaw business and started

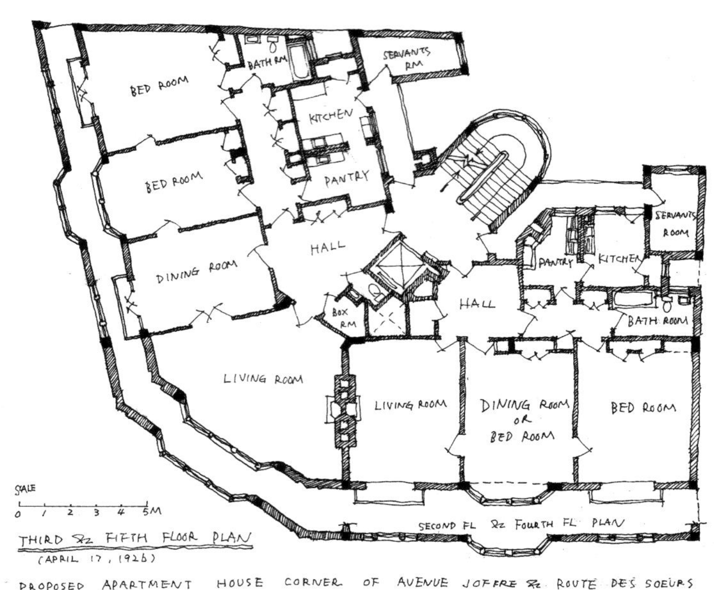

四层、六层平面 3rd Floor Plan and 5th Floor Plan

to develop real estate projects in cooperation with Chinese companies.

Perched at the corner of Avenue Joffre and Route des Coeurs (now Ruijin 1. Road), Estrella Apartment is a seven-story building with 15 suites that serves both commercial and residential purposes.

The ground floor featured a line of shops, each equipped with an independent mini staircase leading up to a two-bedroom suite on the second floor. Every floor above the third floor had two three-bedroom suites with servants' room attached. The two suites in different sizes faced two streets. The apartment was topped by a beautiful roof garden.

Estrella was one of the first apartments equipped with modern facilities including garbage furnace, mechanical refrigerator, wardrobes for different seasons and in-wall ironing board, etc.

Hudec had created smartly designed functional homes for Estrella. The entrance hall was equipped with a mirror and a cabinet to store coat, hat and umbrella. The kitchen was installed with gas or electronic ovens and modern sink. The bathroom contained bow-shaped bathtub and shower enclosure. Even the mirror was illuminated to make shaving more convenient.

The building is detailed in Spanish style with Renaissance and Baroque elements. The elevation is in three sections horizontally. The first and second floors compose the plinth while the third to sixth floors are punctuated with shallow balcony decorated with protruding windows and cast iron patterns. The top of the building is composed of continuous balconies and cornices.

Strong Spanish architectural elements feature all over the building ranging from the patterns on the window lintels, the bamboo-like thin columns set as the glazing bars of the protruding windows, and the geological patterns made of mosaics on the ground and on the wainscot.

The contrasts between creamy walls and the red-brick molding showcase a tone of traditional Eastern European architecture.

The middle part of the Avenue Joffre was also known as "Neva Road of St. Petersberg" since a rainbow of Russian shops had filled the land between Rue Massenet (now Sinan Road) and Shanzhong Road (now Changshu Road). The exotic Estrella Apartments merged perfectly in this backdrop.

Route des Coeurs had been renamed Ruijin 1. Road in 1943 while Estrella also got a new name, the Ruijin Building. It was listed as Shanghai Excellent Historical Buildings for Preservation in 1999.

Tips

The building can be appreciated from the Middle Huaihai Road and Ruijin 1. Road. The interior is not open to the public.

刘吉生住宅 Liu Jisheng's Residence

爱神花园
The Garden of Psyche

因为业主的爱情故事和普绪赫雕像的传奇，位于巨籁达路的刘吉生住宅无疑是邬达克为上海留下的最浪漫的作品。

刘吉生是旧上海人称煤炭大王、火柴大王的著名爱国实业家刘鸿生的胞弟，也是刘氏家族的二号人物。为感谢青梅竹马的贤内助陈定贞家里家外的鼎力支持，1926 年，他邀请邬达克在自己购买的法租界巨籁达路两个地块上设计一座花园洋房，送给妻子作为 40 岁生日礼物。

刘吉生住宅亦称“爱神花园”，因为据说设计受到神话中希腊公主普绪赫和爱神丘比特故事的启发。同样，建筑的柱式风格、花园布局与英国画家莱顿的名作《普绪赫出浴》也颇为相似。

这是一座意大利文艺复兴风格的三层住宅，平面基本为对称布局，但门厅设在东侧。最显眼的是朝南的两层敞廊，由四根通高的爱奥尼柱式支撑，二层平台在柱间微微外鼓，表面的铸铁栏杆纤细而柔美。凉廊正对花园，中轴线上有一座蝶形喷水池，白色大理石的普绪赫雕像从中央升起，双臂高举，逆光而立，格外妩媚。少女的脸微侧，朝向二层东南角的主卧室。

建筑室内不仅装饰华美，而且蕴含着女主人很多巧妙心思。比如，椭圆形主楼梯的铸铁栏杆内嵌入 KSL 字样，那是刘吉生名字的缩写；二楼主卧室的衣橱上玫瑰盛开、蝴蝶翩跹，墙面和天花上也装饰着天使和小小的玫瑰，因为陈定贞的英文名就叫玫瑰。

花园中所有的景观小品，包括石椅、花盆均由邬达克亲自设计后定制加工。花园南面原来是一个网球场，北面沿街则是佣人房。

新中国成立后，刘吉生夫妇移居香港，白头偕老，相守一生。

1952 年后，巨鹿路刘氏住宅成为上海市作家协会的办公楼，原来曾接待过上海滩各路大亨的厅堂成为举办各种文学活动的场所。

最富传奇色彩的是，邬达克为刘陈两人的真挚爱情所打动而专门从意大利定制的普绪赫雕像，在“文革”期间险遭不测，幸得花匠巧妙保护才留存至今。如今这位美丽的爱神之妻已经成为作协的标志，不仅作为纪念品，还以此为书名出版了文集，更有《阮玲玉》等几十部电影、电视剧曾在此取景。

参观指南

私人办公空间，除特别邀请外，不对外开放。

The garden villa on 675 Julu Road is the most romantic of Hudec's work.

Its former owner was Liu Jisheng, a tycoon in Shanghai in the 1920s who, with his elder brother Liu Hongsheng, owned many large enterprises in the city including the Shanghai Cement Company and Hong Kong Matchstick Factory. As a gift on the 40th birthday of his wife Chen Dingzhen (Rose), he bought the site in 1924 to build a house. Rose decided on the

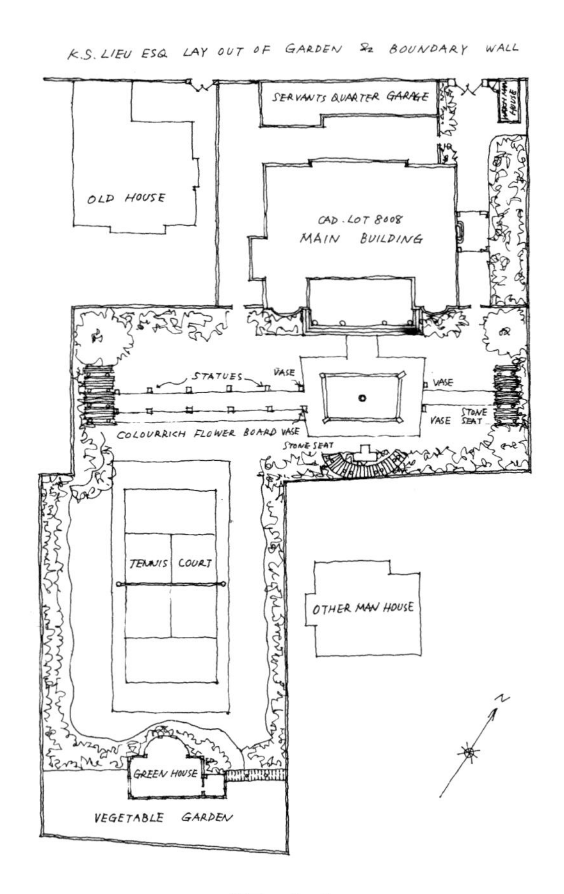

总平面 Site Plan

design and Hudec was the architect.

Today the villa is widely known as "the Garden of Psyche" since the Greek-style house boasts a stunning centerpiece in the garden — a 1.5-meter-high white marble statue of Psyche with four angels and springs of water filling a butterfly-shaped pool beneath her feet. Hudec had got the statue sculpted in Italy as a gift and it was to become the soul of this garden.

It is said Hudec had drawn inspiration from the Greek myth of Eros (the god of love, Cupid in Roman mythology) and Psyche (a mortal, beautiful Greek princess), who fell in love with each other. Even the column styles and garden layout of the villa resemble the masterpiece *"The Bath of Psyche"* by British artist Frederic Leighton.

Hudec himself had designed all details of the garden from stone chairs to even flowerpots. There used to be a tennis court south of the garden.

The three-floor villa is designed in Italian Renaissance style in a symmetrical layout. The entrance hall lies in the eastern façade. The open veranda on the second floor is supported by four gigantic Ionic orders. Between the four orders are three sections of cast iron railings, which are slightly protruding in a soft, delicate pattern.

The first floor features giant halls where large-scale banquets and balls were held. Each pendant lamp hangs from a cloud of brilliant patterns on the ceiling. The second floor was Liu and Chen's bedroom and for VIP guests. The children slept on the third floor. The house with its elaborate garden is redolent with the love and hopes of the woman named "Rose" who lived there 80 years ago. Chen wanted to create a magnificent, smartly designed home to match her husband's booming status and wealth and steady their marriage.

She asked the architect to put "KSL," the abbreviation of her husband's name on the railing of the main staircase. Moreover there are "roses" everywhere in their bedroom. A creamy-hued closet is carved with countless roses and flying angels and tiny roses grace the walls and ceiling. Liu and his wife "Rose" left Shanghai for Hong Kong in 1948 and stayed together until Liu died in 1962. She died two years later and they are buried side-by-side in Montreal, Canada. The villa became the office of the Shanghai Writers' Association after 1952. The former grand hall hosting a rainbow of Shanghai tycoons now hosts many literary events.

And the statue of Psyche, the soul of the garden villa, survived the upheavals of the Cultural Revolution (1966-76) largely due to its former gardener, who buried it under straw in a greenhouse.

Now the statue is back and has become an icon of the writers' association, which not only brought out souvenirs with images of the statue but also published books on it. The beautiful garden has also served as a set for dozens of movies and TV plays.

Tips

The building is not open to the public.

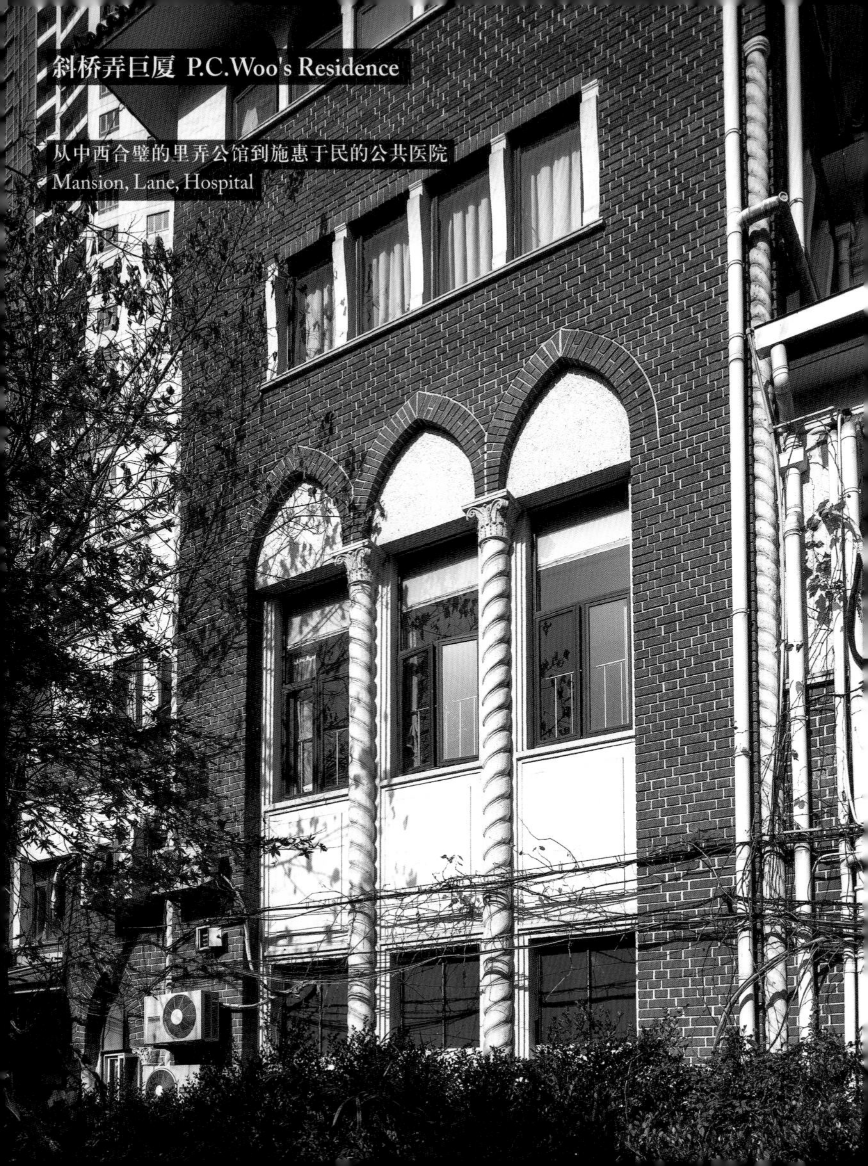

斜桥弄巨厦 P.C.Woo's Residence

从中西合璧的里弄公馆到施惠于民的公共医院
Mansion, Lane, Hospital

出于低调安全等考虑，旧上海很多大户人家喜欢将公馆藏身于里弄之中，斜桥弄巨厦也不例外。这座占地 900 平方米，建筑面积达 2200 平方米的豪宅夹杂在斜桥弄（今吴江路）的多片里弄之间，主人的身份至今仍是未解之谜。虽然设计图上注明这是一位外国人的住宅，但 P.C.WOO 更像一位中国精英的名字缩写，《中国月刊》（1934）曾刊登全套设计图这一点也可作为佐证。

这是一座气势宏大的花园洋房，平面布局呈现中西并置的特点。东面四开间非对称布局，带有门廊、螺旋形大楼梯和车库，每个房间都设壁炉，俨然一副现代化做派。西面首层三开间左右几乎是对称布局，功能沿纵向层层递进：南面为客堂，中间是餐厅和私塾，隔着天井最北面是中式厨房和佣人房，完全沿袭了传统的生活模式和建筑格局。

建筑外形以西班牙风格为主调，兼有其他风格的要素。典型的西班牙建筑特征包括：平缓的筒瓦屋面、螺旋形柱式、南向敞廊和阳台，窗、门廊乃至烟囱顶部均为尖券形。外墙除拉毛粉刷外，还辅以面砖、人造石、斩假石。室内公共部位的地面和柱式均采用大理石，门廊内的柱式和山花为巴洛克风格。

房间内用美国进口的高级橡木地板席纹铺砌。西式古典家具雕刻精美，而堂屋暖气片的铸铁盖板则被塑成各种中国传统吉祥图案，如“五福同祥”、“喜上梅梢”等。宽敞的螺旋形大楼梯每层平台处有一个半圆形凹龛，内置雕塑和一个落地式大西洋钟，楼梯东侧落地长窗饰以彩绘玻璃，室内光线柔和绚丽。

1953 年，斜桥弄巨厦在总工会领导下改为上海市公用事业局职工医院，亦称公惠医院，力主“医疗帮困，施惠于民”。该建筑现为医院三号楼，底层是检验科，二三层则是综合病房。

希望这座设计精美、保存尚好的老房子早日被列入上海市优秀历史建筑保护名录。

参观指南

穿过医院门诊楼可欣赏南立面，室内不对外开放。

科检查
上海市总工会
上海市公惠医院
体检中

P. C. Woo's Residence is a well-designed, well-preserved but little-known work of Hudec near West Nanjing Road but away from the eyes of the public.

This three-story, 2,000-squaremeter building was designed in 1931 to serve as the private residence of a wealthy Chinese family surnamed Woo, whose identity is still a myth.

Today it serves as Gonghui Hospital under the management of the Shanghai Labor Union to serve low-income patients.

Tucked away in a deep lane and hidden behind the clinic hall, this beautiful building was designed in rich Spanish style with added eclectic elements.

Green ivy and artificial green grapes were twining around the red-brick walls that had an air or romance to it. But behind the interesting shapes of the doors and windows, white-uniformed doctors talk to their patients. There is also a thick herbal smell that comes from the traditional Chinese medicine treatment rooms.

The façade was composed of typical Spanish architectural elements, such as Spanish roof tiles, open loggia, spiral columns and cast iron railings. The element of pointed arch repeated on the windows, on the porch and even on the chimney.

The interior was luxurious where East-met-West or it was plain East and West.

The eastern part of this complicated structure was another Western classic that boasted a sitting room, a dining hall, a private study room, Chinese kitchen and servants' quarters.

The centrepiece was a gorgeously curved staircase, which was always stylish like in all Hudec's work. A French window at the east of the staircase was inlayed with stained windows, which introduced a subtle, splendid light into the room.

The flooring on the three floors was of different colors and materials used offered aesthetic variety: black-and-white marble on the first, yellow-and-black terrazzo on the second and black-and-white mosaic tiles on the third. Almost every room had a fireplace.

But the western part of the house contained several very Chinese rooms decorated with stunningly exquisite wooden carvings all over the floors and walls. One room was probably used as a Buddhist shrine before. The iron covers of the heating units were even graced by Chinese lucky patterns, such as "five bats" (meaning five lucky things). Chinese elements were very rare in Hudec's works which may have come from the requests of the Chinese owners.

According to hospital employee Xie Xiaofang, several descendants of the family — all in their 80s — had come down from the US to see their former home in 1995. The descendants recalled their childhood days in this big house such as sliding down the railing of the spiral staircase for fun.

P. C. Woo's Residence is still not a listed historical building, so little attention has been paid to it over the years. And that has made it uncharted and very unique — a building standing in the middle of the garden surrounded by Spanish columns, green ivy and the strong scent of Chinese herbs.

Tips

Since it is hidden inside a lane, it is easy to miss. Remember to cross the clinic hall into the garden to appreciate the south façade. The interior is not open to the public.

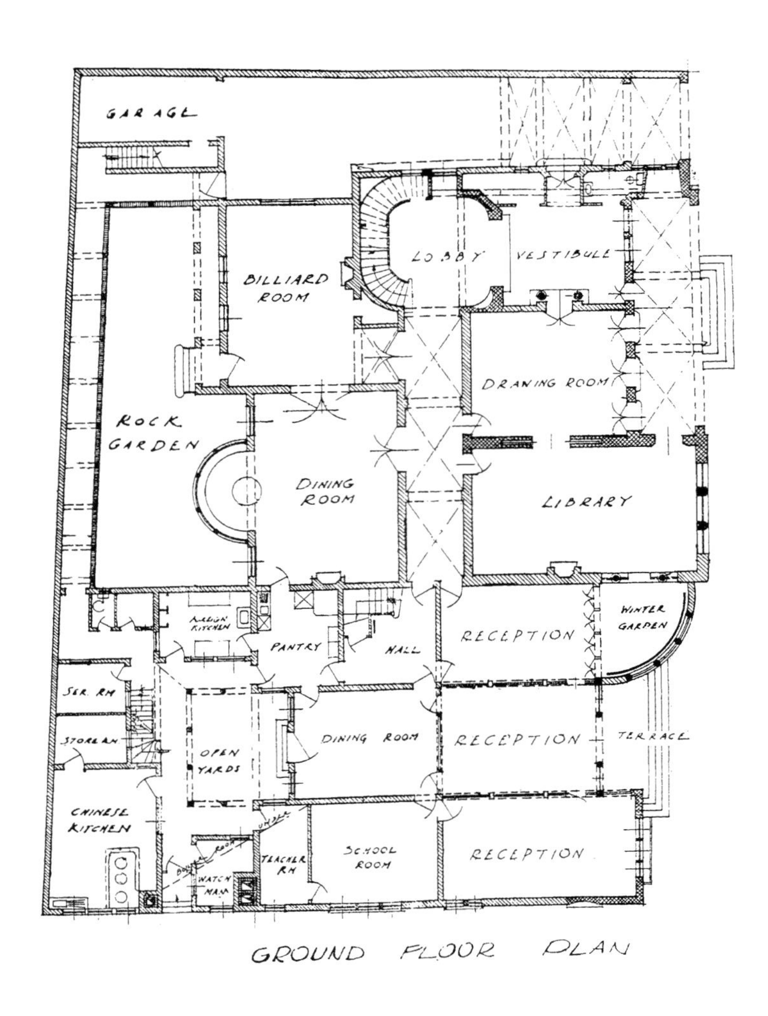

首层平面 Ground Floor Plan

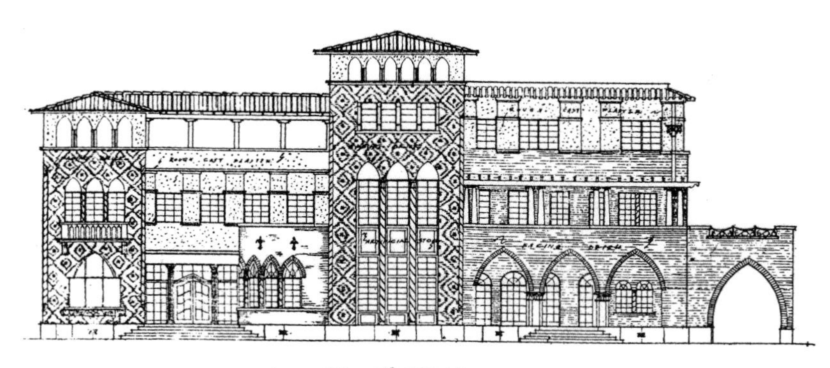

东、南立面 East and South Elevation

震旦女子文理学院
Aurora College for Women

美国圣心女修会在中国创办的唯一女子大学
The Only Women's College Founded by Convent of Sacred Heart in China

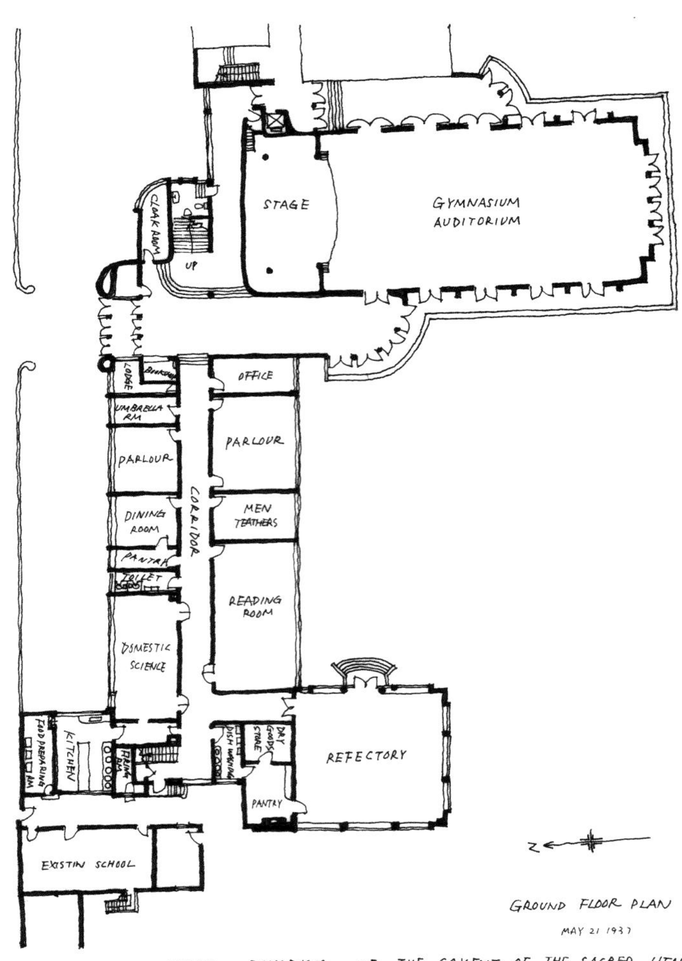

首层平面 Ground Floor Plan

震旦女子文理学院，1937 年 9 月正式开学，名义上挂靠震旦大学并由胡文耀校长兼任院长，实际上属于 1926 年创建的美国圣心女修会，后者是该会在中国创建的天主教区内唯一的女子大学，因此学院的教务长和教导主任都是美国修女，采用英语授课。

震旦女子文理学院包括两部分：东侧是原美国圣心女修会的校舍（现为上海社会科学院出版社办公楼），西侧邬达克设计的新楼与之毗邻，在三四层直接相通。新建筑的平面呈 U 字形，开口向南，东翼有 10 度左右的偏转，与旧楼形成垂直关系。原平面一层用于公共服务，包括体育馆（兼作大会堂）、食堂等，二三层是教室，四层是学生宿舍，屋顶设有活动场地。大会堂上部最初设计成一座教堂，东北角大楼梯上面设一座钟塔，建造时已全部改为教室。

该楼的立面完全是现代主义风格，开长条窗，通高两层的会堂周边是大面积玻璃窗，局部弧形转角的处理为这座完全以功能为导向的教学楼带来了雅致的韵味。新楼于 1939 年 9 月正式启用。

1941 年，著名学者钱钟书先生曾在此任教，杨绛先生的妹妹杨必是其学生。1943－1945 年，校舍被日本占领，成为关押英美等侨民的集中营。1951 年后，该校随震旦大学并入复旦大学和交通大学，原校址与震旦大学附属中学合并成为向明中学。

As the only women's college which was founded by Convent of Sacred Heart in China, Aurora College began its first semester in September 1937. With American nuns as deans and educational directors, the medium of instruction was English.

The college had two buildings. The eastern section was the original school building of the Convent of Sacred Heart, which currently houses the office of Shanghai Social Science Publishing House.

Hudec had designed a new building in the west, which was connected to the eastern building on the third and fourth floors. Shaped like a huge "U," the new building opened to the south, the east wing of which formed a vertical angle to the old one. Originally the ground floor featured a gym, which also operated as an auditorium, besides a dining hall. Classrooms were located on the second and third floors while the fourth floor housed dormitories. The roof had a playing court.

The initial design for a chapel and a bell tower inside the building ended up as classrooms.

The façade boasted a modern style with long windows, grand glass windows and curved corners, giving an elegant look to this functional building. It was put into use in September 1939.

Qian Zhongshu, a famous Chinese literary scholar, taught here in 1941. His students included Yang Bi, sister of his writer wife Yang Jiang. From 1943 to 1945, the campus was occupied by the Japanese as a camp for British and Americans. Along with Aurora University, the college was later merged with Fudan and Jiao Tong universities.

The western building is now part of the Shanghai Xiang Ming High School on Changle Road.

参观指南

沿长乐路可看到北立面，除本校学生外，建筑内部不对外开放。

Tips

The school is not open to the public but the northern façade is visible on the Changle Road.

F 区

8. 中西女塾社交堂(1921-1922)
今市三女中五一大楼
江苏路 155 号(近武定西路)
长宁区

44. 上海啤酒厂(1933-1934)
今苏州河梦清园
宜昌路 130 号(近中潭路)
普陀区

46. 中西女塾景莲堂(1921-1935)
今市三女中五四大楼
江苏路 155 号(近武定西路)
长宁区

Zone F

8. Social Hall, McTyeire School for Girls (1921-1922)
now May 1 Building, Shanghai No.3 Girls High School,
No. 155 Jiangsu Rd (near West Wuding Rd),
Changning District

44. Union Brewery Ltd. (1933-1934)
now Mengqing Park,
No. 130 Yichang Rd (near Zhongtan Rd),
Putuo District

46. McGregor Hall, McTyeire School for Girls (1921-1935)
now May 4 Building, Shanghai No.3 Girls High School,
No. 155 Jiangsu Rd (near West Wuding Rd),
Changning District

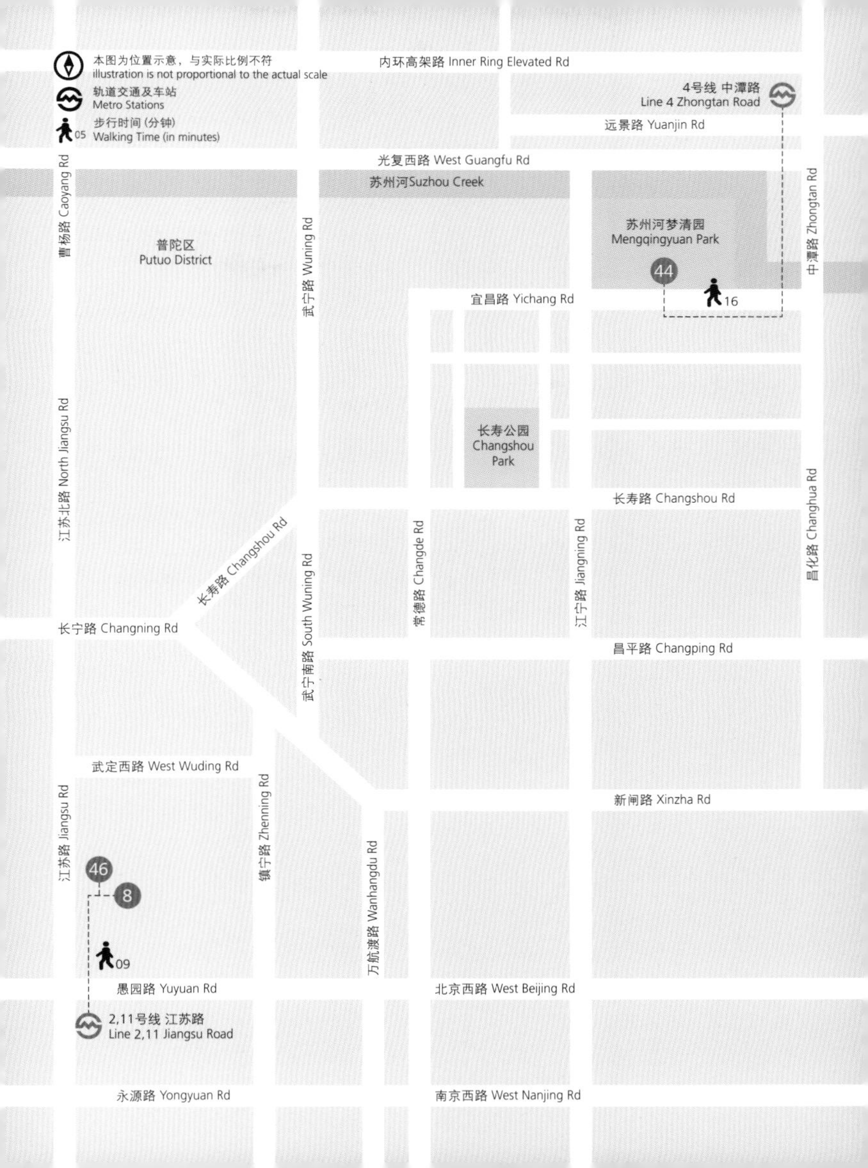
本图为位置示意，与实际比例不符
illustration is not proportional to the actual scale
轨道交通及车站
Metro Stations
步行时间（分钟）
Walking Time (in minutes)
05
内环高架路 Inner Ring Elevated Rd
4号线 中潭路
Line 4 Zhongtan Road
远景路 Yuanjin Rd
光复西路 West Guangfu Rd
苏州河Suzhou Creek
苏州河梦清园
Mengqingyuan Park
44
16
宜昌路 Yichang Rd
曹杨路 Caoyang Rd
普陀区
Putuo District
武宁路 Wuning Rd
中潭路 Zhongtan Rd
长寿公园
Changshou
Park
长寿路 Changshou Rd
江苏北路 North Jiangsu Rd
长寿路 Changshou Rd
武宁南路 South Wuning Rd
常德路 Changde Rd
江宁路 Jiangning Rd
昌化路 Changhua Rd
长宁路 Changning Rd
昌平路 Changping Rd
武定西路 West Wuding Rd
新闸路 Xinzha Rd
江苏路 Jiangsu Rd
镇宁路 Zhenning Rd
万航渡路 Wanhangdu Rd
46
8
09
愚园路 Yuyuan Rd
北京西路 West Beijing Rd
2,11号线 江苏路
Line 2,11 Jiangsu Road
永源路 Yongyuan Rd
南京西路 West Nanjing Rd

中西女塾社交堂
Social Hall, McTyeire School for Girls

中西女塾景莲堂
McGregor Hall, McTyeire School for Girls

中上阶层华人家庭趋之若鹜的淑女学堂
A Girls' School for the Chinese Elites

中西女塾，清光绪十八年（1892）由基督教美国南方监理公会著名传教士林乐知创办，女传教士海淑德任首位校长，英文名是墨梯学校，旨在纪念支持创办该校的墨梯主教。学校初设在汉口路（今沐恩堂东侧，扬子饭店所在地），宋氏三姐妹曾在此校区就读。

虽然最初学生仅 7 名，但出色的教育理念使中西女塾很快成为中上阶层华人家庭趋之若鹜的淑女学堂，规模不断扩大。加之租界地价飞涨，1917 年，学校通过募捐购得忆定盘路（近江苏路）占地 89 亩的经家花园，初二以上年级迁往沪西，小学到初一学生留在原址。

1921 年夏，邬达克代表的克利洋行参加了新校区的建筑设计竞赛，美国学院派哥特风格的精美设计战胜了所有对手，包括金陵女子大学的建筑师亨利·墨菲。

项目初期进度极快，650 名工人同时在现场施工。1922 年春，第一幢高三层半、能容纳 400 名学生的宿舍楼，即社交堂就已建成。大楼平面呈 U 字形，开口朝西，面向花园，底层为平缓的尖券敞廊，高耸的红色双坡瓦屋顶上整齐镶嵌着老虎窗。为纪念刚刚故世的南方监理会主教，该楼取名为蓝华德堂。

此后，因经费短缺，其他建设搁浅。1929 年，中西女塾向中国政府立案，聘杨锡珍为首任中国校长，并于次年更名为私立中西女子中学。直到 1935 年，即邬达克的方案完成 14 年以后，另一幢教学大楼——景莲堂才在北侧建成。该楼高度也是三层半，砖木混合结构，红瓦大坡顶上嵌着老虎窗。平面呈 T 字形，朝南一排是教学楼，背后是一层高的大礼堂。

大楼外观造型简洁，但室内和细部装饰古典唯美。最有特色的是主入口上方三个哥特式尖券窗，白色窗棂，顶部有花瓣饰，内镶几何图案的彩色玻璃，为室内带来温暖柔和的光影。大厅地面是镶嵌图案的彩色磨石子，大礼堂两侧同样采用尖券窗和彩绘玻璃。为了纪念因校园建设过度操劳而早逝的第二任校长，该楼也被称为连吉生堂。

1943 年夏，日军强占中西女中校舍作为陆军第二医院，学校被迫迁出。1952 年，中西女中与圣玛利亚女中（张爱玲母校）合并成为上海市第三女子中学。

今天，邬达克设计的哥特式宿舍楼和教学楼仍优雅地矗立在大草坪的两边，分别成为市三女中的五一大楼和五四大楼，且均已被列入上海市优秀历史建筑保护名录。

参观指南

除本校师生外，建筑不对外开放。

Founded in 1892 by the American missionary Young John Allen who belonged to the Southern Methodist Church, Mctyeier School started off with as few as seven students. However, it soon began to appeal to the Chinese elites who sent their daughters to the school which had by then established a reputation for its superior conceptions on education. Even the famous Soong sisters had briefly studied here on Hankou Road, east of the Moore Memorial Church.

To fulfill the growing needs, the school purchased land on Edinburgh Road (now Jiangsu Road) to house the senior students.

In the summer of 1921 R.A. Curry's firm, represented by Hudec, took part in the bidding to design the new campus buildings. His plan, a finely designed American Gothic building had beaten all the competitors, including American architect Henry Murphy who had built the Ginling College in Nanjing.

With 650 workers, the project took off with great enthusiasm. The social hall, a building as high as three-and-a-half-storeys planned for 400 students, was completed in the spring of 1922. Facing a huge garden, the building featured a lancet corridor and a steep, red-colored double sloping roof graced by dormers. The building was named "Lambuth Hall" in the memory of the late Bishop Water Russell Lambuth.

But shortage of funds led to stoppage of work, meaning the other building, McGregor Hall in the north, was not completed until 1935 — 14 years after Hudec had drafted the plan.

The T-shaped, brick-and-wood structure is also three-and-a-half-storied high, with a giant red-tile sloping roof with dormers. With a line of classrooms facing south and an auditorium in the back, the building revealed a simple style with ancient, elegant architectural details.

Above the main entrance, three white Gothic lancet windows graced by petal patterns and stained glasses, brought warm, subtle lights and shadows to the inside. The floor was paved with patterned, rainbow-hued terrazzo. The building is named "Richardson Hall" after the school's second headmaster Helen Richardson.

The school stopped functioning after the invading Japanese occupied the campus and used it as a military hospital in the summer of 1943. In 1952, McTyeire School became the No. 3 Middle School for girls along with St. Mary's Hall, another top missionary school in the city for girls where the famous writer Aileen Chang had her early education.

The two Hudec buildings, built 14 years apart, have survived the years and serve as the main buildings for the middle school next to a big lawn. They are listed and protected under Shanghai Excellent Historical Buildings.

Tips

The buildings are not open to the public.

上海啤酒厂 Union Brewery Ltd.

中国最大的啤酒加工厂
The Largest Brewery
in China

邬达克在上海建成了两座有影响的工业建筑。一座是华商投资、建在黄浦江畔从苏州河到吴淞口中间的闸北电厂(1930)，另一座就是坐落在苏州河西段的上海啤酒厂。

该工厂1913年由挪威商人创建于江宁路,1931年在香港注册为友啤公司。因为华洋混居时间一长，上海的啤酒消费市场不断扩大，连本地人上餐馆也喜欢点啤酒而不是传统的茶水。在此背景下，友啤公司在公共租界西北角购地筹建大型现代化加工厂。

1931年邬达克在欧洲探亲、旅行，还在德国动了心脏手术，后赴慕尼黑待了很久，仔细研究了啤酒生产工艺。返沪后，设计很快就完成了。

新厂区的总平面依据基地边界布置成马蹄形，主要建筑有：酿造楼、灌装楼、仓库、办公楼和发电间等，建筑面积2.88万平方米。工艺流程全部机械化，所有设备从国外进口，产能为500万瓶啤酒，一度是中国最大的啤酒生产企业。新厂1934年竣工，1936年正式投产。

因为苏州河畔土质疏松，同时又要承受巨大的荷载，比如存放贮藏和发酵罐、啤酒桶等的库房地面荷载为每平方米25吨，这给基础带来了很大困难。最后使用超过2000根木桩来加固地基，深度达到33米。该楼因此成为上海当时最重的建筑。

上海啤酒厂建筑群整体都是从功能出发的现代主义方盒子，全部采用钢筋混凝土结构，但每栋楼体量不一，在细部上也有各自不同的处理。其中灌装楼沿道路呈曲面布置，每层设水平长窗，转角倒圆，底层还采用了无梁楼盖。九层酿造楼体量层层缩进，高度可匹敌当时上海最高的沙逊大厦的屋顶，其东立面和锅炉房的南立面同样采用装饰艺术风格的竖向线条。办公楼的阳台等处也有装饰要素。

1999年，上海啤酒厂被列为上海市优秀历史建筑，岂料不久却险些毁于一旦。2002年,苏州河整治工程完成后，政府拟拆除全部建筑，在此建造大型生态绿地。当规划部门亲临现场中止破坏时，矗立70余年的中国优秀近代工业建筑群只剩下了办公楼、灌装楼、锅炉房，酿造楼也已由九层拆为五层，主体结构开裂，岌岌可危。劫后余生的建筑现为“梦清园”的一部分，灌装楼成为苏州河展示中心，酿造楼局部和锅炉房成为啤酒主题吧，大面积的玻璃幕墙位置就是当初被拆的“伤口”。

参观指南

梦清园开放时间内建筑外观均可参观。苏州河展示中心室内：双休日9:15-16:15，周二、三预约。

Hudec had built two influential industrial buildings in Shanghai. One was the Zhabei Power Station along the Huangpu River with Chinese capital in 1930. The other was the Union Brewery Ltd in the western part of the Suzhou Creek.

The beer factory was initially established on Jiangning Road by Norwegian businessmen in 1913 and registered as "Union Brewery"in Hong Kong in 1931. The cohabitation of Chinese and foreigners had expanded the beer market in Shanghai. Even the locals preferred beer than the

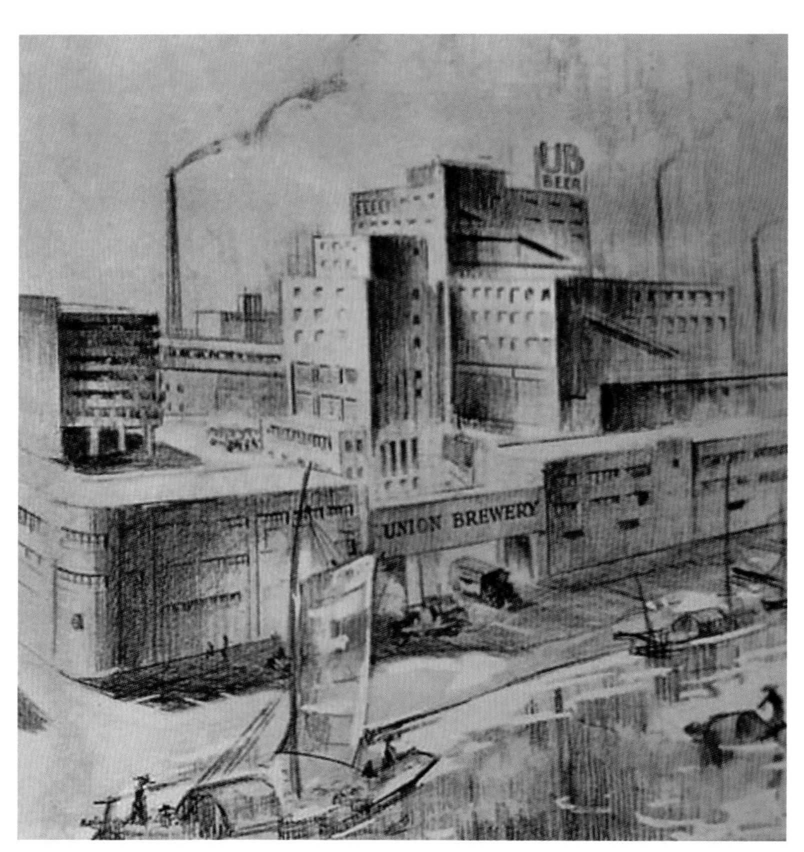

友啤公司沿苏州河透视草图 A sketch of the Union Brewery Ltd. along the Suzhou Creek

traditional Chinese tea when dining out. So the Union Brewery purchased land at the northwest corner of the International Settlement to build a modern beer factory.

Hudec had been travelling in Europe and visiting his family in 1913. He had a heart surgery in Germany and stayed for a long time in Munich, where he carefully studied the process of making beer. So he drafted the plan for the beer factory right after his return to Shanghai.

The general layout of the plant area is designed in the shape of a horseshoe. Covering a building area of 28,800 square meters, the plan contains a brew house, a bottle-filling section, a warehouse, an office building as well as an electric power generation room. All the equipments for the mechanization of techniques were imported from overseas.

Completed in 1934 and put into use in 1936, the brewery used to produce 5 million bottles of beer every year, making it China's largest beer producer.

The average weight for storing the fermentation tanks and beer barrels was up to 25 tons per square meter. It was difficult to build the base for it on the loose soil along the Suzhou Creek. Hudec used more than 2,000 timber piles to strengthen the base, which was as deep as 33 meters. It was the heaviest building in Shanghai at that time.

The brewery is essentially a steel-and-concrete structure and shaped like modern boxes according to their function. But every building differs on the size and architectural details.

The building for beer billing wanders along the road in a curved manner with horizontal long windows on each floor.

The nine-story brewery building shrinks layer upon layer until the top, whose height could compete with the Sassoon House, the highest building in Shanghai at that time. The eastern and the southern façades of the boiler room are emphasized with vertical lines in Art Deco style, which is also repeated on the balcony of the office building.

The Union Brewery was listed as Shanghai Excellent Historical Building in 1999 but was nearly demolished in 2002 when the government planned to build a big botanical greenland in the area after the Suzhou Creek renovation project was completed.

The urban planning department officials intervened and only the office building, the building for beer filling and a boiler room of the 70-year-old factory survived the bulldozers. The nine-story brewery building now has only five stories left with its structure severely damaged.

The surviving buildings have become part of a project named Mengqing Garden, or Garden of a Clean Dream. The brewery building has become an exhibition center for Suzhou Creek, part of which was renovated into a beer-themed bar together with the boiler room. A large glass wall is in fact an "old wound." It was where the building was damaged in 2002.

Tips

The building is open to the public during opening hours of the Mengqing Garden. The Suzhou Creek Exhibition Center is open from 9:15am to 4:15pm on Saturday and Sunday. It is also open on Tuesday and Wednesday but requires reservation.

G 区

31. 息焉堂
   (1929-1931)
   今西郊天主堂
   可乐路 1 号 (近哈密路)
   长宁区

Zone G

31. Sieh Yih Chapel
   (1929-1931)
   now Catholic Country Church
   No. 1 Kele Rd (near Hami Rd),
   Changning District

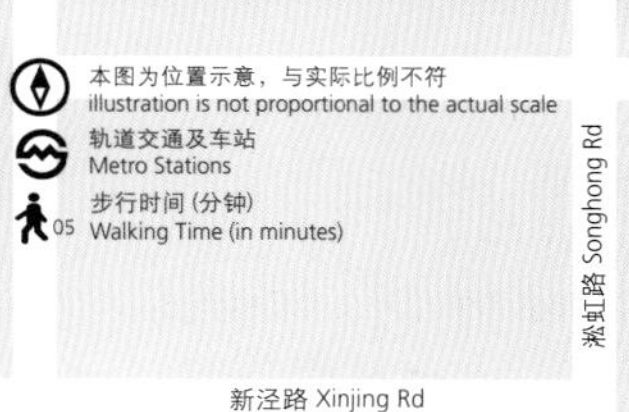

本图为位置示意，与实际比例不符
illustration is not proportional to the actual scale
轨道交通及车站
Metro Stations
05
步行时间（分钟）
Walking Time (in minutes)
可乐路 Kele Rd
31
30
淞虹路 Songhong Rd
哈密路 Hami Rd
新泾路 Xinjing Rd
福泉路 Fuquan Rd
长宁区
Changning District
新潮路 Xinchao Rd
中泾路 ZhongJing Rd
金浜路 Jinbang Rd
剑河路 Jianhe Rd
上海动物园
Shanghai Zoo
清溪路 QingXi Rd
哈密路 Hami Rd
虹桥路 Hongqiao Rd
10号线，上海动物园站
Line 10, Shanghai Zoo

息焉堂 Sieh Yih Chapel

沪上唯一采用拜占庭穹顶
的天主教堂
Shanghai's Only Catholic Church Topped
with a Byzantine Dome

息焉堂是邬达克在上海留下的最具神秘色彩的作品，因为地处偏远西郊，又曾坐落于墓地内，还是唯一采用拜占庭风格穹顶的天主教堂，对于建筑的名称、设计者和建成时间也存在较多争议。

息焉堂全称为息焉公墓圣母升天堂，或罗别根路（今哈密路）西人公墓礼拜堂，为信徒举行安葬仪式祈祷时使用，故亦称安息堂。“息焉”意为“息于此”。公墓的主要发起人，创办震旦、复旦等大学的著名教育家马相伯先生在碑记中点明：选择沪西新泾港干爽地筹建公墓和圣堂，造福教友，仁者在此安息，灵魂升天。1952 年，病逝于越南的若瑟先生被迎回，安葬于此。

同样根据这篇碑记和英文旧报，息焉堂设计时间为 1929 年 5 月，1931 年 8 月 29 日建成开放，而非早先以为的 1925 年。

1929、1933 年数篇报道和草图签名证实，息焉堂确系邬达克作品。1941 年自传中，邬达克解释道，他是上海耶稣会的首席建筑师，虽本人信奉路德教，却曾为不同宗派设计教堂。至于教友中广泛流传的设计者是潘世义，主要原因在于他是公墓五位发起人之一，应该也是中国天主教会的主要建筑师。更大的可能是，他被指派与邬达克合作，负责深化和监理施工。

息焉堂采用早期基督教用于殡葬的纪念性小礼拜堂的单厅式布局，因为那一时期最重视缅怀逝者。平面为正十字形，中厅直接连接东面临河的半圆形圣坛，四角为四根束柱，上承帆拱和穹顶。穹隆底部密排 16 个尖券窗洞，光线漫入时，穹顶宛若飘浮在空中。

教堂外形敦厚，墙面用水泥砂浆作淡黄色鱼鳞纹粉饰，圆顶外表饰以铜板，为留出窗洞折成花瓣形，年久氧化后表面呈蓝绿色。主体西北角设独立钟塔一座，高近 20 米，在各个角度都能形成独特景观。教堂入口和钟塔的窗都采用哥特式尖券。整体建筑采用现代的钢筋混凝土结构。教堂外原是墓区，停尸房等，总占地面积达 4.96 万平方米，逝者可通过地下通道入堂供拜祭。

“文革”初期公墓遭毁，仅存安息堂及钟塔。1974 年，此地成为上海动物园繁殖场物料仓库。2006 年归还教会，整修后于 2008 年 4 月重新启用。

参观指南

建筑室外可供参观，室内每周六 8:00-14:30 对外开放，弥撒时间每周六上午 8:00。

大厅草图 Sketch for Hall

Crouching Tigers and Flying Cranes. In between lies a Byzantine-style Catholic church that is still in service.

Among all the Hudec buildings, none of them is located in an environment as unusual and beautiful as the Catholic Country Church (formerly known as Sieh Yih Chapel) along the Xinjing River in the western suburbs. This is a very rare Byzantine-style church in Shanghai that shows Hudec's mastery of different architectural styles.

Despite some suggestions that Chinese Catholic architect Pan Shiyi was the real man behind the building, recent study has proved beyond doubt that Hudec designed the church with Pan working as his assistant.

It was the first of three churches Hudec designed for Shanghai. In the 1930s, Hudec created the Gothic-style Moore Memorial Church on Xizang Road and the Art Deco New German Lutheran Church, which no longer exists, on Huashan Road.

The church was built in 1929 with funds raised by legendary Chinese Catholic educator Ma Xiangbo and other Catholics. Born in 1840, Ma was a renowned and knowledgeable Catholic Chinese who enjoyed a long life of nearly 100 years. He founded three universities including the prestigious Fudan University.

A visitor can spot the church from a short bridge on Kele Road. Topped with a stunning dome, the church in a stocky shape was reflected perfectly on the tranquil water. The scene was very Zen in some ways.

The original walls, which were coated in rough-textured cement, have been renovated and look less authentic. The most breathtaking part of the building is the dome coated with copper, reveals a cyan hue. Light cast through the window apertures on the bottom of the dome gives the illusion of the huge dome suspended in the air.

Catholic services are still held on Saturdays, most of the participants being elderly, gray-haired Chinese women. During service under the big dome, the old black-and-white pictures on the walls add to the mysterious sanctified atmosphere.

A further northwest of the church, a 20-meter-high bell tower embellished with Gothic windows offers a bird's view of the church and its surroundings.

The church was previously known as the "Chapel of the Westerners Cemetery on Rubicon Road." The 50,000-square-meter open area around the church was primarily used as a cemetery. There was also a mortuary and an underground passage to transfer the remains from the mortuary to the church for the funeral service.

Today the former mortuary is used as the office for the local Catholic community. The building with interesting Chinese elements is preserved in good condition.

Since 1970s, the former cemetery on the west side of the church has become part of the Shanghai Zoo. On the other side of the tower, we can see white cranes patrolling the waterside.

Tips

The church is open from 8am to 2:30pm on Saturdays. Service starts at 8:00am on Saturdays.

邬达克的上海

——这美丽的香格里拉

从1918年11月茫然抵达到1947年1月悄悄逃离，邬达克在上海度过了将近29年光阴，这差不多是他人生一半的时间，而25–54岁正是一个人最年富力强的阶段。来时，他孤身一人，除了战争留下的身心创伤外一无所有；离开时，他拖家带口，带走不少财富和荣誉，留下更多辉煌的建筑遗产。对于这座收留、包容、成就他的异国大都市，邬达克究竟会有怎样的情感？

2010年秋，坐在大光明电影院的金色大厅里，屏幕上年逾古稀的邬达克次子和女儿在上海展开寻根之旅，伴随他们穿街走巷镜头的是当年的老歌——“这美丽的香格里拉，这可爱的香格里拉……我深深地爱上了它，爱上了它……”对了，“这美丽的香格里拉”，用它来描绘上海对于邬达克的意义真是太贴切了！

一位刚刚结束学业，尚未开始事业的欧洲年轻建筑师，被命运之手推到了这座陌生的东方城市。它是一处避难所，它是一处淘金地，但“这美丽的香格里拉”永远无法成为漂泊者的“灵魂之家”。

1 “这美丽的香格里拉”是战争避难所

如果说真有所谓“命运之手”，那么把邬达克从家人身旁推到远隔万水千山的异国他乡，有生之年再无聚首之日的“命运之手”，无疑就是战争。

因为参加“一战”，邬达克成为战俘，被送入西伯利亚的集中营；因为战争未结束，他在被遣送回国时受阻，阴差阳错来到上海；还是因为战争，祖国分裂，父亲遭家乡新的统治者，捷克政府迫害致死，留下巨额的债务和难缠的官司，母亲也郁郁离世，作为长子的邬达克必须留守上海，努力工作，以支撑家庭。

战火第二次燃起时，上海也未能幸免。时任上海匈牙利社区领导人的邬达克，又不惜冒着生命危险保护匈籍犹太人。

上海，“这美丽的香格里拉”是整个邬达克家族，也是其匈牙利同胞的战争避难所。

2 “这美丽的香格里拉”是理想淘金地

20世纪20–40年代，正是上海大兴土木，蓬勃发展的时期。作为一名建筑师，邬达克个人的才华和命运非常幸运地契合了城市发展的良机。凭着扎实的业务和精明的经营，他迅速成为上海滩炙手可热的明星建筑师。他是上海100多栋单体建筑共同的“洋爸爸”，仅仅国际饭店一项就足以使其名垂青史。虽然离沪时，不动产无法带走，但靠他多年在瑞士银行中的存款，邬达克及其家人足以在美国舒适生活，无需再为生计奔波。

上海——“这美丽的香格里拉”还为他带来了很高的社会地位。他是很多高级外国俱乐部的会员，也跟中国的政治、经济和文化精英保持着良好关系，最后还因其显赫社会声望当选匈牙利荣誉领事。

3 “这美丽的香格里拉”并非灵魂归宿

虽然在上海有名有利有地位，有美丽的妻子和活泼的儿女，邬达克应该很幸福。然而事实却表明，他常常紧张焦虑，提心吊胆。这不只是因为公务繁忙，更是因为没有安全感。这种威胁来自“这美丽的香格里拉”所处的残酷现实——有着 20 个仆人的美丽家园建在“路有冻死骨”的战争孤岛中，残酷而脆弱。

距离割断了邬达克跟父母、姊妹和其他亲人之间的联系。战争期间，通信非常困难，他只好通过无线电短波来了解欧洲战事，祈求亲人平安。更糟糕的是，因为奥匈帝国分裂，从 1918 年底开始，邬达克无时无刻不在为争取身份而努力，却屡屡受挫，直到 23 年后才正式拿到匈牙利护照。

跟出生在上海的妻子吉塞拉不同，跟子女更不同，他们是天生的世界公民，家乡、祖国对他们没有太深刻的含义。然而，邬达克是因为躲避战争被迫远离故土亲人的，是有家不能回。虽然在异乡成就辉煌，但内心始终孤独。在晚年写给家人的信中，他称自己为“飞翔的荷兰人”，那位被判罚在鬼船上终生飘泊，直至最后审判日才得以安定的荷兰船长。在美国去世后，妻子遵照其遗愿将他送回拜斯特尔采巴尼亚的家族墓地，与父母亲人团聚。

邬达克终于魂归故里。

László Hudec had spent 29 of his 65 years in Shanghai, from November 1918 to January 1947, from the age of 25 to 54 — the best creative time for a man. He came to the city alone with nothing but a broken heart and an injured leg, both souvenirs of war. And he left the city with wealth, reputation, a family of beautiful wife and three kids and a rainbow of architectural heritage. So what were his true feelings about Shanghai, the exotic city that had received him in his worst days and offered opportunities to great achievements?

During the World Expo year in 2010, a Slovakian documentary on Hudec, "The Man Who Changed Shanghai", was screened at the Golden Cinema Hall of the Grand Theatre, Hudec's signature work. An old Chinese song used in the movie, "The beautiful Shangri-La" was probably the best way to describe the importance of Shanghai in Hudec's life.

A young East European architect was thrown to the strange Oriental city by the hands of destiny. Shanghai was a shelter and a gold mine, but it could never be a "home for the soul" for a vagrant like Hudec.

Shanghai, a shelter from the war

Wars had altered the life of László Hudec forever. This architecture major graduate joined World War I but was taken prisoner by the Russians and sent to Khabarovsk in Siberia from where he escaped and took shelter in Shanghai in 1918.

That same year his home country Austria-Hungary was dissolved after a military defeat on the Italian front in the World War I. Following the war, his father was persecuted to death by the new government, leaving only a huge debt and a bitter lawsuit. His mother passed away soon after. As the eldest son, Hudec had no choice but to stay in Shanghai where he made handsome money and supported his family.

As a Hungarian community leader in Shanghai, Hudec had also risked his life to protect the Hungarian Jews during World War II.

So the "beautiful Shangri-La" Shanghai was truly a shelter for Hudec, his family and his Hungarian compatriots.

Shanghai, a gold mine

Shanghai meanwhile was undergoing massive development from the 1920s to 1940s, which coincided during Hudec's stay in Shanghai. Hudec escaped to the right place at the right time, where innovative ideas and great buildings were mushrooming.

With excellent skills and brilliant management, he soon became a star architect and designed more than 100 buildings including the Park Hotel. Although he could not take the real estate properties away when he left the city in 1947, his savings in the bank of Switzerland guaranteed his family a comfortable life in the US for the rest of his life, with no need to make a living.

Hudec meanwhile had also attained a high social status in Shanghai. As a member of the city's many high-end clubs, he had a good relationship with Chinese elites. Given his influence, he was appointed as an honorary consul of Hungary.

Shanghai, still not a home

With money, fame, a beautiful wife and lovely kids, Hudec should have enjoyed the rest of his life in Shanghai. But there are evidences that he had been "nervous, tense, punctual, even rigid." The reason was not just because of the pressure at work, but also about feeling a lack of security in the city. The beautiful Shangri-La had also some cruel realities — his luxurious home with 20 servants was built in the settlement where people were starving just outside on the streets.

And being far off from his hometown had limited his contacts with his relatives back home. It was often hard to communicate during those war times. He had to rely on radio to keep himself updated about the wars in Europe and pray for his family.

It was made even worse when he lost his nationality after his country was dissolved. Hudec had to wait for 23 years to get a Hungarian passport in 1940.

Hudec's Shanghai-born wife Gizella and his kids were all "cosmopolitans," to whom hometown or homeland did not mean much. But Hudec suffered. He was driven away from home by the war and then had a home but could never return. Despite his huge success in a foreign land, Hudec suffered from bouts of loneliness. In a letter to a family member, he wrote "Wherever I go I would be a foreigner or a guest, a Flying Dutchman who has a home everywhere and yet has no home or a house anywhere."

After he died, his wife respected his will to send him back to his family's vault in Besztercebanya (now in Slovakia) to join his parents and relatives.

Only then Hudec finally returned home.

上海邬达克建筑不完全名录

Shanghai Hudec Architecture Incomplete Directory

1 巨籁达路 22 栋住宅（1919-1920）
巨鹿路 852 弄 1-10 号、巨鹿路 868-892 号
（近常熟路）
静安区

2 中华懋业银行上海分行（1919-1920）
原南京路 11 号，已拆除
黄浦区

3 美丰银行（1920）
今美丰大楼
河南中路 521-529 号（近宁波路）
黄浦区

4 何东住宅（1919-1920）
今商务办公楼
陕西北路 457 号（近北京西路）
静安区

5 盘滕住宅（1919-1920）
今上海沪剧院（改造中）
汾阳路 150 号（近桃江路）
徐汇区

6 梅里霭（曼）住宅（1921）
今花园住宅
桃江路 25 号
徐汇区

7 霍肯多夫住宅（1921）
今花园住宅
淮海中路 1893 号
徐汇区

2

1 The 22 Residences on Route Ratard
(1919-1920)
No.1-10 Lane 852 Julu Rd,
No.868-892 Julu Rd (near Changshu Rd)
Jing'an District

2 Chinese-American Bank of Commerce
(1919-1920)
former No.11 Nanking Rd, demolished
Huangpu District

3 American Oriental Banking Corporation (1920)
now Meifeng Building
No. 521-529 Middle He'nan Rd
(Near Ningbo Rd)
Huangpu District

4 Ho Tung's Residence (1919-1920)
now "The Masion" Office Building
No. 457 North Shaanxi Rd
(near West Beijing Rd)
Jing'an District

5 Jean Beudin's Residence (1919-1920)
now Shanghai Shanghai Opera House
(Under Renovation)
No.150 Fenyang Rd (near Taojiang Rd)
Xuhui District

6 Meyrier or Merriman Residence (1921)
now Garden Villa
No. 25 Taojiang Rd
Xuhui District

7 Huckendoff Residence (1921)
now Garden Villa
No.1893 Middle Huaihai Rd
Xuhui District

8 中西女塾社交堂（1921-1922）
今市三女中五一大楼
江苏路 155 号（近武定西路）
长宁区

9 马迪耶住宅（1922）
今上海工艺美术博物馆
汾阳路 79 号
徐汇区

10 卡尔登剧院（1923）
原长江剧院，1993 年拆除
黄河路 21 号
黄浦区

11 美国总会（1922-1924）
原高法大楼
福州路 209 号（近河南中路）
黄浦区

12 方西马大楼（1924-1925）
原上海医药公司，1997 年因建延安东路高架拆除
延安东路 7 号（近外滩）
黄浦区

13 诺曼底公寓（1923-1926）
今武康大楼
淮海中路 1836-1858 号（近武康路）
徐汇区

14 邬达克首座自宅（1922-1926）
原吕西纳路（今利西路）17 号，已拆除
长宁区

15 宏恩医院（1923-1926）
今华东医院 1 号楼
延安西路 221 号（近乌鲁木齐北路）
静安区

12 14

8 Social Hall, McTyeire School for Girls (1921-1922)
now May 1 Building, Shanghai No.3 Girls High School
No. 155 Jiangsu Rd (near West Wuding Rd)
Changning District

9 H. Madier Residence (1922)
now Shanghai Museum of Arts and Crafts
No. 79 Fenyang Rd
Xuhui District

10 Carlton Theatre (1923)
former Changjiang Theatre, demolished in 1993
No. 21 Huanghe Rd
Huangpu District

11 American Club (1922-1924)
former Shanghai High People's Court
No. 209 Fuzhou Rd (near Middle He'nan Rd)
Huangpu District

12 Foncim Building (1924-1925)
former Shanghai Medical Building, demolished in 1997
No. 7 East Yan'an Rd (near the Bund)
Huangpu District

13 Normandie Apartments (1923-1926)
now Wukang Building
No. 1836-1858 Middle Huaihai Rd (near Wukang Rd)
Xuhui District

14 Hudec's 1st Residence (1922-1926)
former No.17 Lucerne Rd (now Lixi Road), demolished
Changning District

15 Country Hospital (1923-1926)
now No. 1 Building of Huadong Hospital
No. 221 West Yan'an Rd (near North Wulumuqi Rd)
Jing'an District

16 宝隆医院（1925-1926）
原长征医院特诊楼，已拆除
凤阳路 415 号
黄浦区

17 普益地产公司巨福路花园住宅（1925-1926）
今花园住宅
乌鲁木齐南路 153、154、155、160、180、182 号
徐汇区

18 爱司公寓（1926-1927）
今瑞金大楼
瑞金一路 148-150 号（近淮海中路）
黄浦区

19 派克路机动车库（1927）
今国际饭店附楼
黄河路凤阳路口
黄浦区

20 四行储蓄会联合大楼（1926-1928）
今联合大楼
四川中路 261 号（近汉口路）
黄浦区

21 西门外妇孺医院（1926-1928）
今上海医科大学附属妇产科医院
方斜路 419 号
黄浦区

22 普益地产公司西爱咸斯路花园住宅（1925-1930）
今花园住宅
安亭路 41 弄 16、18 号，81 弄 2、4 号；永嘉路 563、615、623 号
徐汇区

16 Paulun Hospital (1925-1926)
former Changzheng Hospital, demolished
No. 415 Fengyang Rd
Huangpu District

17 Garden Villas on Route Dufour for Asia Realty Co. (1925-1926)
now Garden Villas
No. 153, 154, 155, 160, 180, 182 South Wulumuqi Rd
Xuhui District

18 Estrella Apartments (1926-1927)
now Ruijin Building
No. 148-150 Ruijin 1. Rd
(near Middle Huaihai Rd)
Huangpu District

19 New Garage & Service Station of Messrs. Honigsberg Co. (1927)
now Skirt Building of Park Hotel
Fengyang Rd and Huanghe Rd
Huangpu District

20 Union Building of the Joint Savings Society (1926-1928)
now Union Building
No. 261 Middle Sichuan Rd (near Hankou Rd)
Huangpu District

21 Margaret Williamson Hospital (1926-1928)
now Maternity Hospital
No. 419 Fangxie Rd
Huangpu District

22 Garden Villas on Route Sieyes for Asia Realty Co. (1925-1930)
now Garden Villas
No. 16, 18 Lane 41, No 2,4 Lane 81 Anting Rd， No. 563, 615, 623 Yongjia Road
Xuhui District

16 19 21

23 西爱威斯路外国人私宅（1929-1930）
今上海往事餐厅
永嘉路 628 号
徐汇区

24 德利那齐宅（1929-1930）
今花园住宅
武康路 129 号
徐汇区

25 闸北电厂（1930）
今上海闸北发电厂
军工路 4000 号
杨浦区

26 浙江电影院（1929-1930）
今浙江电影院
浙江中路 123 号
黄浦区

27 邬达克自宅（1930）
今邬达克纪念室（底层）
番禺路 129 号（近平武路）
长宁区

28 虹桥路雷文住宅（1930）
今龙柏饭店 3 号楼
虹桥路 2419 号
长宁区

29 刘吉生住宅（1926-1931）
今上海市作家协会
巨鹿路 675 号（近陕西南路）
静安区

30 孙科住宅（1929-1931）
今"万科之家"（改造中）
延安西路 1262 号（近番禺路）
长宁区

23 H. Vladimiroff Residence on Route Sieyes (1929-1930)
now Shang Hai Wang Shi Restaurant
No. 628 Yongjia Rd
Xuhui District

24 D. Tirinnanzi Residence on Route Ferguson (1929-1930)
now Garden Villa
No. 129 Wukang Rd
Xuhui District

25 Chapei Power Station (1930)
now Shanghai Chapei Power Station
No. 4000, Jungong Rd
Yangpu District

26 Chekiang Cinema (1929-1930)
now Zhejiang Cinema
No. 123 Middle Zhejiang Rd
Huangpu District

27 Hudec's Residence (1930)
now Hudec Memorial Hall (ground floor)
No. 129 Panyu Rd (near Pingwu Rd)
Changning District

28 Cottage for Frank Raven (1930)
now Cypress Hotel, Villa No.3
No.2419 Hongqiao Rd
Changning District

29 Liu Jisheng's Residence (1926-1931)
now Shanghai Writers' Association
No. 675 Julu Rd (near South Shaanxi Rd)
Jing'an District

30 Sun Ke's Residence (1929-1931)
now Shanghai Vanke House (Under Renovation)
No. 1262 West Yan'an Rd (near Panyu Rd)
Changning District